Rainer Ansorge and Thomas Sonar

Mathematical Models of Fluid Dynamics

For additional information reagarding this topic, please refer also to the following publications

Durbin, P. A., Reif, B. A. P.

Statistical Theory and Modeling for Turbulent Flows

2001
ISBN 978-0-471-49744-8

Shivamoggi, B. K.

Theoretical Fluid Dynamics

1998
ISBN 978-0-471-05659-1

Middleman, S.

Introduction to Fluid Mechanics
Principles of Analysis and Design

1998
ISBN 978-0-471-18209-2

Rainer Ansorge and Thomas Sonar

Mathematical Models of Fluid Dynamics

Modeling, Theory, Basic Numerical Facts
An Introduction

Second, Updated Edition

WILEY-VCH

WILEY-VCH Verlag GmbH & Co. KGaA

The Authors

Prof. Dr. Rainer Ansorge
Norderstedt, Germany
ansorge@math.uni-hamburg.de

Prof. Dr. Thomas Sonar
Institut Computational Mathematics
Braunschweig, Germany
t.sonar@tu-bs.de

Library of Congress Card No.: applied for
British Library Cataloguing-in-Publication Data: A catalogue record for this book is available from the British Library.
Bibliographic information published by the Deutsche Nationalbibliothek
The Deutsche Nationalbibliothek lists this publication in the Deutsche Nationalbibliografie; detailed bibliographic data are available on the Internet at ⟨http://dnb.d-nb.de⟩.

Typesetting le-tex publishing services oHG, Leipzig
Printing Strauss GmbH, Mörlenbach
Binding Litges & Dopf GmbH, Heppenheim
Cover Spiesz Design, Neu-Ulm

Printed in the Federal Republic of Germany
Printed on acid-free paper

ISBN 978-3-527-40774-3

Dedicated to our family members and to our students

Contents

Preface to the Second Edition *XI*

Preface to the First Edition *XIII*

1 Ideal Fluids *1*
1.1 Modeling by Euler's Equations *1*
1.2 Characteristics and Singularities *10*
1.3 Potential Flows and (Dynamic) Buoyancy *14*
1.4 Motionless Fluids and Sound Propagation *29*

2 Weak Solutions of Conservation Laws *33*
2.1 Generalization of What Will Be Called a Solution *33*
2.2 Traffic Flow Example with Loss of Uniqueness *37*
2.3 The Rankine–Hugoniot Condition *42*

3 Entropy Conditions *49*
3.1 Entropy in the Case of an Ideal Fluid *49*
3.2 Generalization of the Entropy Condition *53*
3.3 Uniqueness of Entropy Solutions *59*
3.4 Kruzkov's Ansatz *69*

4 The Riemann Problem *73*
4.1 Numerical Importance of the Riemann Problem *73*
4.2 The Riemann Problem for Linear Systems *75*
4.3 The Aw–Rascle Traffic Flow Model *77*

5 Real Fluids *79*
5.1 The Navier–Stokes Equations Model *79*
5.2 Drag Force and the Hagen–Poiseuille Law *85*
5.3 Stokes Approximation and Artificial Time *90*
5.4 Foundations of the Boundary Layer Theory and Flow Separation *95*
5.5 Stability of Laminar Flows *102*

Mathematical Models of Fluid Dynamics. R. Ansorge and T. Sonar

ISBN: 978-3-527-40774-3

5.6 Heated Real Gas Flows *104*
5.7 Tunnel Fires *106*

6 Proving the Existence of Entropy Solutions by Discretization Procedures *113*
6.1 Some Historical Remarks *113*
6.2 Reduction to Properties of Operator Sequences *114*
6.3 Convergence Theorems *117*
6.4 Example *120*

7 Types of Discretization Principles *127*
7.1 Some General Remarks *127*
7.2 Finite Difference Calculus *131*
7.3 The CFL Condition *135*
7.4 Lax–Richtmyer Theory *136*
7.5 The von Neumann Stability Criterion *141*
7.6 The Modified Equation *144*
7.7 Difference Schemes in Conservation Form *146*
7.8 The Finite Volume Method on Unstructured Grids *148*
7.9 Continuous Convergence of Relations *151*

8 A Closer Look at Discrete Models *155*
8.1 The Viscosity Form *155*
8.2 The Incremental Form *156*
8.3 Relations *158*
8.4 Godunov Is Just Good Enough *159*
8.5 The Lax–Friedrichs Scheme *164*
8.6 A Glimpse of Gas Dynamics *168*
8.7 Elementary Waves *171*
8.8 The Complete Solution to the Riemann Problem *178*
8.9 The Godunov Scheme in Gas Dynamics *184*

9 Discrete Models on Curvilinear Grids *187*
9.1 Mappings *187*
9.2 Transformation Relations *190*
9.3 Metric Tensors *192*
9.4 Transforming Conservation Laws *193*
9.5 Good Practice *196*
9.6 Remarks Concerning Adaptation *203*

10 Finite Volume Models *205*
10.1 Difference Methods on Unstructured Grids *205*
10.2 Order of Accuracy and Basic Discretization *208*
10.3 Higher-Order Finite Volume Schemes *209*
10.4 Polynomial Recovery *211*

10.5 Remarks Concerning Non-polynomial Recovery *216*
10.6 Remarks Concerning Grid Generation *218*

Index *221*

Suggested Reading *227*

Preface to the Second Edition

Because the first edition of this introduction to a mathematical description of fluid dynamics was so enthusiastically adopted by readers, the publishers encouraged the first author to revise the first edition and add any new results and considerations that are relevant to an introductory course. Because of this, the first author introduced some new sections and remarks (e.g., Sections 4.3 , 5.6 , 5.7 , 7.9, etc.), and in turn encouraged the second author (who had already written parts of Chapter 7 for the first edition) to widen his participation in the book and to contribute additional chapters touching upon some modern areas of the numerical treatment of fluid flows to this new edition. However, even in these new chapters, the introductory character of the book has been maintained.

For more extensive representations of special areas of mathematical fluid dynamics, like modeling, theory or numerical methods, we refer the reader to some monographs listed at the end of this book.

Again, we thank the publishers for their support, our colleagues for fruitful discussions, and Mrs. Monika Jampert for technical help with respect to some LaTeX problems. We also hope that readers of the new edition will again consider it useful for enriching their scientific interests and for executing their work.

Hamburg and Brunswick, September 2008

Rainer Ansorge
Thomas Sonar

Mathematical Models of Fluid Dynamics. R. Ansorge and T. Sonar

ISBN: 978-3-527-40774-3

Preface to the First Edition

Mathematical modeling is the process of representing problems from fields beyond mathematics itself using mathematics. The subsequent mathematical treatment of this model using theoretical and/or numerical procedures proceeds as follows:

1. Transition from the nonmathematical phenomenon to a mathematical description, which at the same time leads to the translation of problems formulated in terms of the original problem into mathematical problems.

   This task forces the scientist or engineer who intends to use mathematical tools to:
   - Cooperate with experts working in the field that the original problem comes from. Thus, he/she has to learn the language of these experts and to understand their way of thinking (teamwork).
   - Create or accept an idealized description of the original phenomena, i.e., to ignore the properties of the original problem that are expected to be of no great relevance to the questions under consideration. These simplifications are useful since they reduce the complexity of both the model and its mathematical treatment.
   - Identify structures within the idealized problem and replace these structures with suitable mathematical structures.

2. Treatment of the mathematical substitute.

   This task normally requires:
   - Independent activity from the person working on the problem, who must work theoretically.
   - Treatment of the problem using tools from mathematical theory.
   - The solution of the particular mathematical problems that occur using these theoretical tools; in other words, differential equations or integral equations, optimal control problems, or systems of algebraic equations, etc., must be solved. Numerical procedures are often the only way to do this and to answer the particular questions of interest, at least approximately. The error in the approximate solution compared with the unknown

Mathematical Models of Fluid Dynamics. R. Ansorge and T. Sonar

ISBN: 978-3-527-40774-3

so-called exact solution does not normally affect the answer to the original problem a great deal, provided that the numerical method and tools applied are of sufficiently high accuracy. In this context, it should be realized that the quest for an exact solution does not make sense because of the idealizations mentioned above, and because the initial data presented with the original problem normally originate from statistics or from experimental measurements.

3. Retranslation of the results.

   The qualitative and quantitative statements obtained from the mathematical model then need to be retranslated into the language in which the original problem was formulated. In other words, the results must be interpreted with respect to their real-world meaning. This process again requires teamwork with the experts from the field in which the problem originated.

4. Model checking.

   After retranslation, the results must be checked for relevance and accuracy, e.g., by performing experimental measurements. This work must be done by the experts from the field in which the problem originated. If the mathematical results coincide sufficiently well with the results from experiments stimulated by the theoretical forecasts, the mathematical part is then completed, and a new tool that can be applied by the physicists, engineers, etc., to similar situations has been created.

   On the other hand, if the logical or computational errors are nontrivial, the model must be revised. In this situation, the gap between the results from the mathematical model and the real results can only have originated from using too much idealization during the modeling process.

The development of mathematical models not only stimulates new experiments and leads to constructive prognostic (and hence technical) tools for physicists or engineers, but it is also important from the point of view of the theory of cognition: it allows us to understand the connections between different elements from an unstructured set of observations; in other words, to create theories.

Mathematical descriptions have been used for centuries in various fields, such as physics, engineering, music, etc. Mathematical models are also used in modern biology, medicine, philology and economics, as well as in certain fields of art, like architecture or oriental ornaments.

This book presents an introduction to models used in fluid mechanics. Important properties of fluid flows can be derived theoretically from such models. We discuss some basic ideas for the construction of effective numerical procedures. Hence, all aspects of theoretical fluid dynamics are addressed: modeling, mathematical theory and numerical methods.

We do not expect the reader to be familiar with a lot of experimental work. A knowledge of some fundamental principles of physics, like conservation of mass, conservation of energy, etc., is sufficient. The most important idealization is (in contrast to the molecular structure of materials) the assumption of fluid continua.

The reader will find the mathematics in the text easier to understand if he/she is acquainted with some basic elements of:

- Linear algebra
- Calculus
- Partial differential equations
- Numerical analysis
- The theory of complex functions
- Functional analysis

Functional analysis only plays a role in the somewhat general theory of discretization algorithms described in Chapter 6. In this chapter, the question of the existence of weak entropy solutions of the problems under consideration is discussed. Physicists and engineers are normally not very interested in the treatment of this problem. Nevertheless, we felt that it should be included in this work so that this question not left unanswered. The lack of a solution immediately shows that a model does not fit reality if there is a measurable course of physical events. Existence theorems are therefore important beyond the field of mathematics alone. However, readers who are unacquainted with functional analytic terminology can of course skip this chapter.

With respect to models and their theoretical treatment, as well as to the numerical procedures that occur in Sections 4.1, 5.3, 6.3 and Chapter 7, a brief introduction to mathematical fluid mechanics as provided by this book can only present the most basic facts. However, the author hopes that this overview will generate interest in this field among young scientists, and that it will familiarize people working in institutes and industry with some fundamental mathematical aspects.

Finally, I wish to thank several colleagues for suggestions, particularly Thomas Sonar, who contributed to Chapter 7 when we organized a joint course for graduate students,[1] and Dr. Michael Breuss, who read the manuscript carefully. Last but not least, I thank the publishers, especially Dr. Alexander Grossmann, for their encouragement.

1) Parts of Chapters 1 and 5 are translations from parts of Sections 25.1, 25.2, 29.9 of: R. Ansorge, H. Oberle: *Mathematik für Ingenieure, vol. 2, 2nd ed.* Berlin: Wiley-VCH 2000.

1
Ideal Fluids

1.1
Modeling by Euler's Equations

Physical laws are mainly derived from conservation principles, such as conservation of mass, conservation of momentum, and conservation of energy.

Let us consider a fluid (gas or liquid) in motion, i.e., the flow of a fluid.[1] Let

$$\boldsymbol{u}(x, y, z, t) = \begin{pmatrix} u_1(x, y, z, t) \\ u_2(x, y, z, t) \\ u_3(x, y, z, t) \end{pmatrix}$$

be the velocity,[2] and denote by $\varrho = \varrho(x, y, z, t)$ the density of this fluid at point $\boldsymbol{x} = (x, y, z)$ and at time instant t.

Let us take out of the fluid at a particular instant t an arbitrary portion of volume $W(t)$ with surface $\partial W(t)$. The particles of the fluid now move, and assume that $W(t + h)$ is the volume formed at the instant $t + h$ by the same particles that formed $W(t)$ at time t.

Moreover, let $\varphi = \varphi(x, y, z, t)$ be one of the functions describing a particular state of the fluid at time t at point $\boldsymbol{x}$, such as mass per unit volume (= density), interior energy per volume, momentum per volume, etc. Hence, $\int_{W(t)} \varphi \, \mathrm{d}(x, y, z)$ gives the full amount of mass or interior energy, momentum, etc., of the volume $W(t)$ under consideration.

We would like to find the change in $\int_{W(t)} \varphi \, \mathrm{d}(x, y, z)$ with respect to time, i.e.,

$$\frac{\mathrm{d}}{\mathrm{d}t} \int\limits_{W(t)} (x, y, z, t) \, \mathrm{d}(x, y, z) \ . \tag{1.1}$$

1) Flows of other materials can be included too, e.g., the flow of cars on highways, provided that the density of cars or particles is sufficiently high.

2) Note that bold letters in equations normally indicate vectors or matrices.

Mathematical Models of Fluid Dynamics. R. Ansorge and T. Sonar

ISBN: 978-3-527-40774-3

We have

$$\frac{\mathrm{d}}{\mathrm{d}t}\int\limits_{W(t)} \varphi(x,y,z,t)\,\mathrm{d}(x,y,z) = \lim_{h\to 0}\frac{1}{h}\left\{\int\limits_{W(t+h)} \varphi(\tilde{y},t+h)\,\mathrm{d}(y_1,y_2,y_3) - \int\limits_{W(t)} \varphi(x,y,z,t)\,\mathrm{d}(x,y,z)\right\},$$

where the change from $W(t)$ to $W(t+h)$ is obviously given by the mapping

$$\tilde{y} = \boldsymbol{x} + h\cdot \boldsymbol{u}(\boldsymbol{x},t) + \boldsymbol{o}(h)$$
$$\left(\tilde{y} = (y_1,y_2,y_3)^T\right).$$

The error term $\boldsymbol{o}(h)$ also depends on $\boldsymbol{x}$ but the property $\lim_{h\to 0}\frac{1}{h}\boldsymbol{o}(h) = 0$ if differentiated with respect to space, provided that these spatial derivatives are bounded.

The transformation of the integral taken over the volume $W(t+h)$ to an integral over $W(t)$ by substitution requires the integrand to be multiplied by the determinant of this mapping, i.e., by

$$\begin{vmatrix} (1+h\,\partial_x u_1) & h\,\partial_y u_1 & h\,\partial_z u_1 \\ h\,\partial_x u_2 & (1+h\,\partial_y u_2) & h\,\partial_z u_2 \\ h\,\partial_x u_3 & h\,\partial_y u_3 & (1+h\,\partial_z u_3) \end{vmatrix} + o(h)$$
$$= 1 + h\cdot\left(\partial_x u_1 + \partial_y u_2 + \partial_z u_3\right) + o(h)$$
$$= 1 + h\cdot \operatorname{div}\boldsymbol{u}(\boldsymbol{x},t) + o(h)\ .$$

Taylor expansion of $V\varphi(\tilde{y},t+h)$ around $(\boldsymbol{x},t)$ therefore leads to

$$\frac{\mathrm{d}}{\mathrm{d}t}\int\limits_{W(t)} \varphi(x,y,z,t)\ \mathrm{d}(x,y,z) = \int\limits_{W(t)} \left\{\partial_t\varphi + \varphi\ \operatorname{div}\boldsymbol{u} + \langle \boldsymbol{u},\nabla\varphi\rangle\right\}\ \mathrm{d}(x,y,z)\ . \quad (1.2)$$

Here, ∇v denotes the gradient of a scalar function v, and $\langle\cdot\ ,\ \cdot\rangle$ means the standard scalar product of two vectors out of $\mathbb{R}^3$.

The product rule from differentiation gives:

$$\varphi\operatorname{div}\boldsymbol{u} + \langle \boldsymbol{u},\nabla\varphi\rangle = \operatorname{div}(\varphi\cdot\boldsymbol{u})\ ,$$

so that (1.2) leads to the so-called *Reynolds' transport theorem*[3)]

$$\frac{\mathrm{d}}{\mathrm{d}t}\int\limits_{W(t)} \varphi(x,y,z,t)\ \mathrm{d}(x,y,z) = \int\limits_{W(t)} \left\{\partial_t\varphi + \operatorname{div}(\varphi\boldsymbol{u})\right\}\ \mathrm{d}(x,y,z)\ . \quad (1.3)$$

As already mentioned, the dynamics of fluids can be described directly by conservation principles and – as far as gases are concerned – by an additional equation of state.

3) Osborne Reynolds (1842–1912); Manchester

1. Conservation of mass: If there are no sources or losses of fluid within the subdomain of the flow under consideration, the mass remains constant.

Because $W(t)$ and $W(t+h)$ consist of the same particles, they have the same mass. The mass of $W(t)$ is given by $\int_{W(t)} \varrho(x, y, z, t)\, \mathrm{d}(x, y, z)$, and therefore

$$\frac{\mathrm{d}}{\mathrm{d}t} \int\limits_{W(t)} \varrho(x, y, z, t)\, \mathrm{d}(x, y, z) = 0$$

must hold. Taking (1.3) into account (particularly for $\varphi = \varrho$), this leads to the requirement

$$\int\limits_{W(t)} \{\partial_t \varrho + \operatorname{div}(\varrho \boldsymbol{u})\}\, \mathrm{d}(x, y, z) = 0\,.$$

Since this has to hold for arbitrary $W(t)$, the integrand has to vanish:

$$\boxed{\partial_t \varrho + \operatorname{div}(\varrho \boldsymbol{u}) = 0\,.} \tag{1.4}$$

This equation is called the *continuity equation.*

2. Conservation of momentum: Another conservation principle concerns the momentum of a mass, which is defined as

mass × velocity .

Thus,

$$\int\limits_{W(t)} \varrho \boldsymbol{u}\, \mathrm{d}(x, y, z)$$

gives the momentum of the mass at time t of the volume $W(t)$ and

$$\boldsymbol{q} = \begin{pmatrix} q_1 \\ q_2 \\ q_3 \end{pmatrix} = \varrho \boldsymbol{u}$$

describes the *density of momentum.*

The principle of the conservation of momentum, i.e., Newton's second law

force = mass × acceleration,

then states that the change of momentum with respect to time equals the sum of all of the exterior forces acting on the mass of $W(t)$.

In order to describe these exterior forces, we take into account that there is a certain pressure $p(\boldsymbol{x}, t)$ at each point $\boldsymbol{x}$ in the fluid at each instant t. If $\boldsymbol{n}$ is considered to be the unit vector normal on the surface $\partial W(t)$ of $W(t)$, and it is directed outwards, the fluid outside of $W(t)$ acts on $W(t)$ with a force given by

$$-\int_{\partial W(t)} p\,\boldsymbol{n}\,\mathrm{do} \quad (\mathrm{do} = \text{area element of } \partial W(t)) \;.$$

Besides the normal forces per unit surface area generated by the pressure, there are also tangential forces which act on the surface due to the friction generated by exterior particles along the surface.

Though this so-called fluid *viscosity* leads to a lot of remarkable phenomena, we are going to neglect this property at the first step. Instead of *real fluids* or *viscous fluids*, we restrict ourselves in this chapter to so-called *ideal fluids* or *inviscid fluids*. This restriction to ideal fluids, particularly to *ideal gases*, is one of the idealizations mentioned in the Preface.

However, as well as exterior forces per unit surface area, there are also exterior forces per unit volume – e.g., the weight.

Let us denote these forces per unit volume by $\boldsymbol{k}$, such that Newton's second law leads to

$$\frac{\mathrm{d}}{\mathrm{d}t}\int_{W(t)} \boldsymbol{q}\,\mathrm{d}(x,y,z) = \int_{W(t)} \boldsymbol{k}(x,y,z,t)\,\mathrm{d}(x,y,z) - \int_{\partial W(t)} p\cdot\boldsymbol{n}\,\mathrm{do} \;.$$

Thus, by Gauss' divergence theorem, we find

$$\int_{\partial W(t)} p\,\boldsymbol{n}\,\mathrm{do} = \begin{pmatrix} \int_{\partial W(t)} p\,n_1\,\mathrm{do} \\ \int_{\partial W(t)} p\,n_2\,\mathrm{do} \\ \int_{\partial W(t)} p\,n_3\,\mathrm{do} \end{pmatrix} = \begin{pmatrix} \int_{W(t)} \partial_x p\,\mathrm{d}(x,y,z) \\ \int_{W(t)} \partial_y p\,\mathrm{d}(x,y,z) \\ \int_{W(t)} \partial_z p\,\mathrm{d}(x,y,z) \end{pmatrix}$$

$$= \int_{W(t)} \nabla p\,\mathrm{d}(x,y,z) \;.$$

Together with (1.3),

$$\int_{W(t)} \left\{ \partial_t \boldsymbol{q} + \begin{pmatrix} \mathrm{div}(q_1\boldsymbol{u}) \\ \mathrm{div}(q_2\boldsymbol{u}) \\ \mathrm{div}(q_3\boldsymbol{u}) \end{pmatrix} - \boldsymbol{k} + \nabla p \right\} \mathrm{d}(x,y,z) = 0$$

follows.

Again, this has to be valid for any arbitrarily chosen volume $W(t)$. If, moreover,

$$\mathrm{div}(q_i\,\boldsymbol{u}) = \langle \boldsymbol{u}, \nabla q_i \rangle + \mathrm{div}\,\boldsymbol{u}\cdot q_i$$

is taken into account,

$$\partial_t \boldsymbol{q} + \langle \boldsymbol{u}, \nabla \rangle\; \boldsymbol{q} + \mathrm{div}\,\boldsymbol{u}\cdot\boldsymbol{q} + \nabla p = \boldsymbol{k} \;,$$

i.e.,

$$\boxed{\partial_t q + \frac{1}{\varrho}\,\langle q, \nabla\rangle\, q + \operatorname{div}\left(\frac{1}{\varrho}\, q\right)\, q + \nabla p = k} \tag{1.5}$$

has to be fulfilled.

The number of equations represented by (1.5) equals the spatial dimension of the flow, i.e., the number of components of $\boldsymbol{q}$ or $\boldsymbol{u}$.

By means of the continuity equation, (1.5) can be reformulated as

$$\boxed{\partial_t \boldsymbol{u} + \langle \boldsymbol{u}, \nabla\rangle\, \boldsymbol{u} + \frac{1}{\varrho}\, \nabla p = \hat{\boldsymbol{k}}\,,} \tag{1.6}$$

where the force $\boldsymbol{k}$ per unit volume has been replaced by the force $\hat{\boldsymbol{k}} = \frac{1}{\varrho}\,\boldsymbol{k}$ per unit mass.

Equation (1.4) together with (1.5) or (1.6) are called *Euler's equations.*[4)]

$$\varrho,\ E,\ q_1,\ q_2,\ q_3$$

are sometimes called *conservative variables* whereas $\varrho,\ \varepsilon,\ u_1,\ u_2,\ u_3$ are the *primitive variables.* Here,

$$E := \varrho\,\varepsilon + \frac{\varrho}{2}\,\|\boldsymbol{u}\|^2 = \varrho\,\varepsilon + \frac{\|\boldsymbol{q}\|^2}{2\varrho}$$

gives the *total energy* per unit volume, where ε stands for the *interior energy* per unit mass, e.g., the heat per unit mass.

$$\frac{\varrho}{2}\,\|\boldsymbol{u}\|^2$$

obviously introduces the *kinetic energy* per unit volume.[5)]

$\langle \boldsymbol{u}, \nabla\rangle\, \boldsymbol{u}$ is called the *convection term.*

Remark

Terms of the form $\partial_t \boldsymbol{w} + \langle \boldsymbol{w}, \nabla\rangle\, \boldsymbol{w}$ are often abbreviated in the literature to $\frac{D\boldsymbol{w}}{Dt}$ and are called *material time derivatives* of the vector-valued function $\boldsymbol{w}$.

3. Conservation of energy: Next we consider the *first law of thermodynamics*, namely:

The change per time unit in the total energy of the mass of a moving fluid volume equals the work done per time unit against the exterior forces.

4) Leonhard Euler (1707–1783); Basel, Berlin, St. Petersburg

5) $\|\cdot\|$: 2-norm

For ideal fluids, this means that

$$\frac{\mathrm{d}}{\mathrm{d}t} \int\limits_{W(t)} E(x, y, z, t)\, \mathrm{d}(x, y, z) = \int\limits_{W(t)} \left\langle \varrho\, \hat{\boldsymbol{k}}, \boldsymbol{u} \right\rangle \mathrm{d}(x, y, z) - \int\limits_{\partial W(t)} \langle p\, \boldsymbol{n}, \boldsymbol{u} \rangle \, \mathrm{do}\,.$$ [6]

The relation

$$\int\limits_{\partial W(t)} \langle p\, \boldsymbol{n}, \boldsymbol{u} \rangle \, \mathrm{do} = \int\limits_{\partial W(t)} \langle p\, \boldsymbol{u}, \boldsymbol{n} \rangle \, \mathrm{do} = \int\limits_{W(t)} \mathrm{div}(p\, \boldsymbol{u})\, d(x, y, z)$$

follows from Gauss' divergence theorem, so that (1.3) leads to

$$\int\limits_{W(t)} \left\{ \partial_t E + \mathrm{div}(E \cdot \boldsymbol{u}) + \mathrm{div}(p \cdot \boldsymbol{u}) - \left\langle \varrho\, \hat{\boldsymbol{k}}, \boldsymbol{u} \right\rangle \right\} \mathrm{d}(x, y, z) = 0\,, \quad \forall\, W(t)\,.$$

If $\boldsymbol{k}$ can be neglected because of the small weight of the gas, or if $\hat{\boldsymbol{k}}$ is the weight of the fluid per unit mass[7] and $\boldsymbol{u}$ is the velocity of flow parallel to the Earth's surface, we get

$$\boxed{\partial_t E + \mathrm{div}\left(\frac{E + p}{\varrho}\, \boldsymbol{q} \right) = 0\,.} \tag{1.7}$$

Explicitly written, and neglecting $\boldsymbol{k}$, Eqs. (1.4), (1.5) and (1.7) become

$$\begin{aligned}
&\partial_t \varrho + \partial_x q_1 + \partial_y q_2 + \partial_z q_3 = 0 \\
&\partial_t q_1 + \partial_x \left(\frac{1}{\varrho}\, q_1 q_1 + p \right) + \partial_y \left(\frac{1}{\varrho}\, q_1 q_2 \right) + \partial_z \left(\frac{1}{\varrho}\, q_1 q_3 \right) = 0 \\
&\partial_t q_2 + \partial_x \left(\frac{1}{\varrho}\, q_2 q_1 \right) + \partial_y \left(\frac{1}{\varrho}\, q_2 q_2 + p \right) + \partial_z \left(\frac{1}{\varrho}\, q_2 q_3 \right) = 0 \\
&\partial_t q_3 + \partial_x \left(\frac{1}{\varrho}\, q_3 q_1 \right) + \partial_y \left(\frac{1}{\varrho}\, q_3 q_2 \right) + \partial_z \left(\frac{1}{\varrho}\, q_3 q_3 + p \right) = 0 \\
&\partial_t E + \partial_x \left(\frac{E + p}{\varrho}\, q_1 \right) + \partial_y \left(\frac{E + p}{\varrho}\, q_2 \right) + \partial_z \left(\frac{E + p}{\varrho}\, q_3 \right) = 0\,.
\end{aligned} \tag{1.8}$$

Hence, we are concerned with a system of equations that can be used to determine the functions ϱ, q_1, q_2, q_3, E. However, we must note that there is an additional function that is sought, namely the pressure p.

In the case of constant density[8] ϱ, i.e., $\partial_t \varrho = 0 \;\; \forall\, (x, y, z)$, only four conservation variables have to be determined, so the five equations in (1.8) are sufficient. Otherwise, particularly in the case of gas flow, a sixth equation is needed, namely an

6) Work = force × length = pressure × area × length $\Rightarrow \frac{\text{work}}{\text{time}}$ = force × velocity = pressure × area × velocity

7) In other words, $\|\hat{\boldsymbol{k}}\| = g$ where g is the Earth's gravitational acceleration.

8) The case of constant density is not necessarily identical to the case of an *incompressible* flow defined by $\mathrm{div}\, \boldsymbol{u} = 0$, because in this case the continuity equation (1.4) is already fulfilled if the pair $(\varrho \boldsymbol{u})$ shows the property $\partial_t \varrho + \langle \boldsymbol{u} \boldsymbol{\nabla} \varrho \rangle = 0$, which does not necessarily imply $\varrho = \text{const.}$

equation of state. State variables of a gas are:

$$T = \text{temperature}\,, \quad p = \text{pressure}\,, \quad \varrho = \text{density}\,, \quad V = \text{volume}\,,$$
$$\varepsilon = \text{energy/mass}\,, \quad S = \text{entropy/mass}\,,$$

and a theorem of thermodynamics says that each of these state variables can be uniquely expressed in terms of two of the other state variables. Such relations between three state variables are called *equations of state*. Thus, p can be expressed by ϱ and ε (and hence by $\varrho, E, \boldsymbol{q}$); for inviscid so-called γ*-gases*, the relation is given by:

$$p = (\gamma - 1)\,\varrho\varepsilon = (\gamma - 1)\left(E - \frac{\|\boldsymbol{q}\|^2}{2\varrho}\right) \tag{1.9}$$

with $\gamma = \text{const} > 1$, such that only the functions ϱ, $\boldsymbol{q}$, E have to be determined. Here, the *adiabatic exponent* γ is the ratio $\frac{c_p}{c_V}$ of the specific heats. In the case of air, we have $\gamma \approx 1.4$.

Using vector-valued functions, systems of differential equations can also be described by a single differential equation. In this way, and taking (1.9) into account as an additional equation, (1.8) can be written as

$$\partial_t\, V + \partial_x f_1(V) + \partial_y f_2(V) + \partial_z f_3(V) = 0 \tag{1.10}$$

with

$$V = (\varrho, q_1, q_2, q_3, E)^T$$

and with

$$f_1(V) = \begin{pmatrix} q_1 \\ \frac{1}{\varrho}\, q_1 q_1 + p \\ \frac{1}{\varrho}\, q_2 q_1 \\ \frac{1}{\varrho}\, q_3 q_1 \\ \frac{E+p}{\varrho}\, q_1 \end{pmatrix}, \quad f_2(V) = \begin{pmatrix} q_2 \\ \frac{1}{\varrho}\, q_2 q_1 \\ \frac{1}{\varrho}\, q_2 q_2 + p \\ \frac{1}{\varrho}\, q_3 q_2 \\ \frac{E+p}{\varrho}\, q_2 \end{pmatrix}, \quad f_3(V) = \begin{pmatrix} q_3 \\ \frac{1}{\varrho}\, q_3 q_1 \\ \frac{1}{\varrho}\, q_3 q_2 \\ \frac{1}{\varrho}\, q_3 q_3 + p \\ \frac{E+p}{\varrho}\, q_3 \end{pmatrix}.$$

The functions $f_j(V)$ are called *fluxes*.

If Jf_1, Jf_2, Jf_3 are the *Jacobians*, (1.10) becomes

$$\partial_t\, V + Jf_1(V) \cdot \partial_x\, V + Jf_2(V) \cdot \partial_y\, V + Jf_3(V) \cdot \partial_z\, V = 0\,. \tag{1.11}$$

Because of their particular meaning in physics, systems of differential equations of type (1.10) are called systems of *conservation laws*, even if they do not arise from physical aspects. Obviously, (1.11) is a quasilinear system of first-order partial differential equations.

Such systems must be defined by initial conditions

$$V(\boldsymbol{x}, 0) = V_0(\boldsymbol{x})\,, \tag{1.12}$$

where $V_0(\boldsymbol{x})$ is a prescribed initial state, and by boundary conditions.

If the flow does not depend on time, it is called a *stationary* or steady-state flow, and initial conditions do not occur. Otherwise, the flow is termed *nonstationary*.

As far as ideal fluids are concerned, it can be assumed that the fluid flow is tangential along the surface of a solid body[9] fixed in space or along the bank of a river, etc. In this case,

$$\langle \boldsymbol{u}, \boldsymbol{n} \rangle = 0 \tag{1.13}$$

is one of the boundary conditions, where $\boldsymbol{n}$ are the outward-directed normal unit vectors along the surface of the body.

There are often symmetries with respect to space such that the number of unknowns can be reduced, e.g., in the case of rotational symmetry combined with polar coordinates. If it is found that there is only one spatial coordinate, the problem is termed *one-dimensional*.[10] If rectangular space variables are used and x is the only one that remains, we end up with the system

$$\partial_t V + Jf(V) \cdot \partial_x V = 0 \tag{1.14}$$

with

$$V = \begin{pmatrix} \varrho \\ q \\ E \end{pmatrix} , \quad f(V) = \begin{pmatrix} q \\ \frac{1}{\varrho} q^2 + p \\ \frac{E+p}{\varrho} q \end{pmatrix}$$

and with the equation of state

$$p = (\gamma - 1) \left(E - \frac{q^2}{2\varrho} \right) .$$

Hence,

$$Jf(V) = \begin{pmatrix} 0 & 1 & 0 \\ -\frac{3-\gamma}{2} \frac{q^2}{\varrho^2} & (3-\gamma) \frac{q}{\varrho} & \gamma - 1 \\ (\gamma - 1) \frac{q^3}{\varrho^3} - \gamma \frac{Eq}{\varrho^2} & \gamma \frac{E}{\varrho} - \frac{3(\gamma-1)}{2} \frac{q^2}{\varrho^2} & \gamma \frac{q}{\varrho} \end{pmatrix} .$$

As can easily be verified, $\lambda_1 = \frac{q}{\varrho}$ is a solution of the characteristic equation of $Jf(V)$, namely of

$$-\lambda^3 + 3 \frac{q}{\varrho} \lambda^2 - \left\{ (\gamma^2 - \gamma + 6) \frac{q^2}{2\varrho^2} - \gamma(\gamma - 1) \frac{E}{\varrho} \right\} \lambda + \left(\frac{\gamma^2}{2} - \frac{\gamma}{2} + 1 \right) \frac{q^3}{\varrho^3} - \gamma(\gamma - 1) \frac{Eq}{\varrho^2} = 0 .$$

9) For example, a wing

10) Two- or three-dimensional problems are defined analogously.

The other two eigenvalues of $Jf(V)$ are then the roots of

$$\lambda^2 - \frac{2q}{\varrho}\lambda + (\gamma^2-\gamma+2)\frac{q^2}{2\varrho^2} - \gamma(\gamma-1)\frac{E}{\varrho} = 0:$$

$$\begin{aligned}\lambda_{2,3} &= \frac{q}{\varrho} \pm \sqrt{\frac{q^2}{\varrho^2} + \gamma(\gamma-1)\frac{E}{\varrho} - (\gamma^2-\gamma+2)\frac{q^2}{2\varrho^2}} \\ &= \frac{q}{\varrho} \pm \sqrt{\gamma(\gamma-1)\frac{E}{\varrho} - \gamma(\gamma-1)\frac{q^2}{2\varrho^2}} \\ &= \frac{q}{\varrho} \pm \sqrt{\frac{\gamma(\gamma-1)}{\varrho}\left[E - \frac{q^2}{2\varrho}\right]} = \frac{q}{\varrho} \pm \sqrt{\gamma\frac{p}{\varrho}}\,.\end{aligned} \tag{1.15}$$

Thus, these eigenvalues are real and different from each other ($p > 0$).

Definition

If all the eigenvalues of a matrix $\boldsymbol{A}(x,t,\boldsymbol{V})$ are real and if the matrix can be diagonalized, the system of equations

$$\partial_t \boldsymbol{V} + \boldsymbol{A}(x,t,\boldsymbol{V})\,\partial_x \boldsymbol{V} = 0$$

is termed *hyperbolic* at (x,t,V). If the eigenvalues are real and different from each other such that $\boldsymbol{A}$ can definitely be diagonalized, the system is said to be *strictly hyperbolic.*

Obviously then, (1.14) is strictly hyperbolic for all (x,t,V) under consideration.

By the way, because $\frac{q}{\varrho} = u$, the velocity of the flow, and because $\sqrt{\frac{\gamma p}{\varrho}}$ describes the local sound velocity $\hat{c}$ (cf. (1.66)), the eigenvalues are:

$$\lambda_1 = u\,, \quad \lambda_2 = u + \hat{c}\,, \quad \lambda_3 = u - \hat{c}\,,$$

and have equal signs in the case of supersonic flow, whereas the subsonic flow is characterized by the fact that one eigenvalue has a different sign to the others.

Definition

Let $\boldsymbol{u}(\boldsymbol{x},t_0)$ be the velocity field of the flow at instant t_0 and let $\boldsymbol{x}_0$ be an arbitrary point from the particular subset of $\mathbb{R}^3$ which is occupied by the fluid at this particular instant. If, at this moment, the system

$$\left[\boldsymbol{x}'(s), \boldsymbol{u}(\boldsymbol{x}(s), t_0)\right] = 0^{11)}$$

of ordinary differential equations with the initial condition

$$\boldsymbol{x}(0) = \boldsymbol{x}_0$$

has a unique solution $\boldsymbol{x} = \boldsymbol{x}(s)$ for each of these points $\boldsymbol{x}_0$, each of the curves $\boldsymbol{x} = \boldsymbol{x}(s)$ is called a *streamline* at instant t_0.

Here, the streamlines may be parametrized by the arc length s with $s = 0$ at point x_0, such that

$$\langle x', x' \rangle = 1 \, .$$

Thus, the set of streamlines at an instant t_0 shows a snapshot of the flow at this particular instant. It does not necessarily describe the trajectories along which the fluid particles move over time. Only if the flow is a stationary one, i.e., if $\boldsymbol{u}$, ϱ, p, $\hat{\boldsymbol{k}}$ are independent of time, do the streamlines and trajectories coincide: a particle moves along a fixed streamline over time.

1.2
Characteristics and Singularities

As an introductory example for a more general investigation of conservation laws, let us consider the scalar *Burgers' equation*[12] without exterior forces:

$$\partial_t v + \partial_x \left(\frac{1}{2} v^2 \right) = 0 \, , \quad (x, t) \in \Omega \, , \quad \text{i.e.,} \quad x \in \mathbb{R} \, , \quad t \geq 0 \, , \tag{1.16}$$

which is often studied as a model problem from a theoretical point of view and also as a test problem for numerical procedures.

Here, the flux is given by $f(v) = \frac{1}{2}v^2$. If

$$v_0(x) = 1 - \frac{x}{2} \tag{1.16a}$$

is chosen as a particular example of an initial condition, the *unique* and *smooth solution*[13] turns out to be

$$v(x, t) = \frac{2 - x}{2 - t} \, ,$$

but this solution only exists locally, namely for $0 \leq t < 2$; as time increases it runs into a singularity at $t = 2$.

As a matter of fact, classical existence and uniqueness theorems for quasilinear first-order partial differential equations with smooth coefficients only ensure the unique existence of a classical smooth solution in a certain neighborhood of the initial manifold, provided that the initial condition is also sufficiently smooth.

The occurrence of discontinuities or singularities does not depend on the smoothness of the fluxes: assume that $v(x, t)$ is a smooth solution of the problem

$$\begin{aligned} &\partial_t v + \partial_x f(v) = 0 \quad \text{for} \quad x \in \mathbb{R} \, , \quad t \geq 0 \\ &v(x, 0) = v_0(x) \end{aligned}$$

11) Here, $[\cdot, \cdot]$ is the vector product in $\mathbb{R}^3$.

12) J. Burgers: Nederl. Akad. van Wetenschappen 43 (1940) 2–12.

13) In other words, v is continuously differentiable.

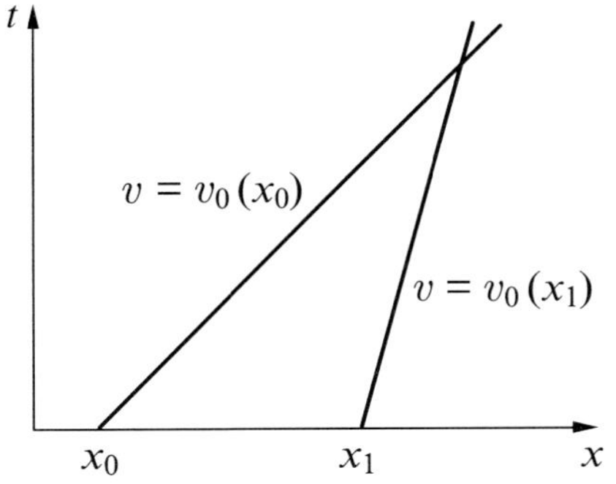

Fig. 1.1 Formation of discontinuities.

in a certain neighborhood immediately above the x-axis. Obviously, this solution is constant along the straight line

$$x(t) = x_0 + tf'(v_0(x_0)) \tag{1.17}$$

that crosses the x-axis at x_0, where $x_0 \in \mathbb{R}$ is chosen arbitrarily. Its value along this line is therefore $v(x(t), t) = v_0(x_0)$. This can easily be verified by

$$\begin{aligned}\frac{\mathrm{d}}{\mathrm{d}t}v(x(t), t) &= \partial_t v(x(t), t) + \partial_x v(x(t), t) \cdot x'(t) \\ &= \partial_t v(x, t) + \partial_x v(x, t) \cdot f'(v_0(x_0)) \\ &= \partial_t v(x, t) + \partial_x v(x, t) \cdot f'(v(x, t)) = \partial_t v + \partial_x f(v) = 0\ .\end{aligned}$$

The straight lines of (1.17), each of which belong to a particular x_0, are called the *characteristics* of the given conservation law.

In the case of $x_0 < x_1$, but for example $0 < f'(v_0(x_1)) < f'(v_0(x_0))$,[14] the characteristic through $(x_0, 0)$ intersects the characteristic through $(x_1, 0)$ at an instant $t_1 > 0$, so that at the point of intersection the solution v must have the value $v_0(x_0)$ as well as the value $v_0(x_1) \neq v_0(x_0)$. Therefore, a discontinuity will occur at the instant t_1 or even earlier. With respect to fluid dynamics, one notable type of discontinuity is shocks (discussed later).

When applied to Burgers' equation (1.16) with an initial function of (1.16a), the characteristic through a point x_0 on the x-axis is given by

$$t = 2\,\frac{x - x_0}{2 - x_0}\ ,$$

such that the discontinuity we found at $t = 2$ can also be immediately understood via Fig. 1.2.

If systems of conservation laws are considered instead of the scalar case, i.e.,

$$V(x, t) \in \mathbb{R}^m\ ,\ m \in \mathbb{N}\ ,\ \forall\,(x, t) \in \Omega\ ,$$

and if only one spatial variable x occurs, a system of characteristics is defined as follows:

14) Hence, $v_0(x_0) \neq v_0(x_1)$ too

Definition

If $V(x, t)$ is a solution of (1.10), in the case of only one space variable x, the one-parameter set

$$x_{(i)} = x_{(i)}(t, \chi) \quad (\chi \in \mathbb{R} : \text{set parameter})$$

of real curves defined by the ordinary differential equation

$$\dot{x} = \lambda_i V(x, t) \tag{1.18}$$

for every fixed i ($i = 1, \ldots, m$) is called the set of *i-characteristics* of the particular system that belongs to V. Here, $\lambda_i V)$ ($i = 1, \cdots, m$) are the eigenvalues of the Jacobian $\boldsymbol{Jf}(\boldsymbol{V})$.

Obviously, this definition coincides in the case of $m = 1$ with the previously presented definition of characteristics.

Let us finally – using an example – study the situation for a system of conservation laws if the system is linear with constant coefficients:

$$\partial_t \boldsymbol{V} + \boldsymbol{A}\,\partial_x \boldsymbol{V} = 0 \tag{1.19}$$

with a constant Jacobian (m, m)-matrix $\boldsymbol{A}$. Moreover, let us assume the system to be strictly hyperbolic.

From (1.18), the characteristics turn out to be the set of straight lines given by

$$x_{(i)}(t) = \lambda_i t + x_{(i)}(0) , \quad (i = 1, \ldots, m) ,$$

and are independent of $\boldsymbol{V}$. Hence, for every fixed i, the characteristics belonging to this set are parallel.

If $\boldsymbol{S} = (s_1, s_2, \ldots, s_m)$ is the matrix whose columns consist of the eigenvectors of the Jacobian, and if $\boldsymbol{\Lambda}$ denotes the diagonal matrix consisting of the eigenvalues of $\boldsymbol{A}$,

$$\boldsymbol{A} = \boldsymbol{S}\boldsymbol{\Lambda}\,\boldsymbol{S}^{-1} \tag{1.20}$$

follows.

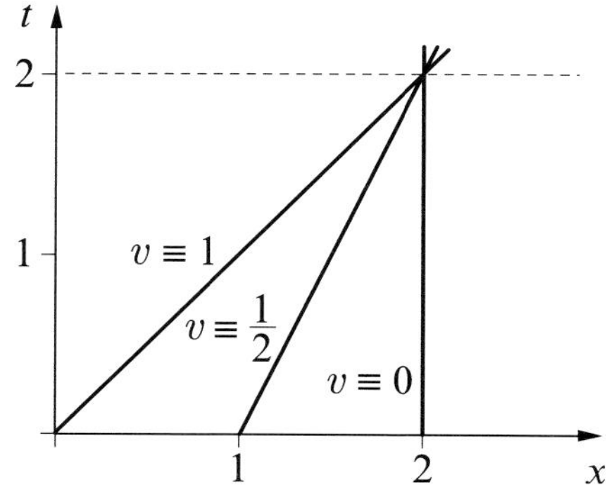

Fig. 1.2 Discontinuity of the solution to Burgers' equation.

If new variables $\hat{V}$ are introduced by

$$\hat{V} = S^{-1} V ,$$

the system takes the form

$$\partial_t \hat{V} + \Lambda\, \partial_x \hat{V} = 0 , \quad \hat{V}_0 = S^{-1} V_0 . \tag{1.21}$$

This is a decoupled system:

$$\partial_t \hat{v}_i + \lambda_i\, \partial_x \hat{v}_i = 0 , \; \hat{v}_i(x, 0) = \left[S^{-1} V_0(x)\right]_i = \hat{v}_{i_0}(x) , \quad (i = 1, \ldots, m) .$$

Each of the equations of this system is an independent scalar equation called an *advection equation*. The solution is

$$\hat{v}_i(x, t) = \hat{v}_{i_0}(x - \lambda_i\, t) \quad (i = 1, \ldots, m) . \tag{1.22}$$

Hence, the state at instant t moves with velocity λ_i in the positive or negative x-direction according to the sign of λ_i. This is called wave propagation, where the velocity of propagation is described by λ_i.

Obviously,

$$V = (s_1, \ldots, s_m) \begin{pmatrix} \hat{v}_1 \\ \vdots \\ \hat{v}_m \end{pmatrix} = \hat{v}_1 s_1 + \hat{v}_2 s_2 + \ldots + \hat{v}_m\, s_m .$$

Each of the vector-valued functions

$$V_{(i)}(x, t) := \hat{v}_i(x, t) \cdot s_i$$

solves the system of differential equations because of

$$\begin{aligned} \partial_t V_{(i)} + S \Lambda S^{-1}\, \partial_x V_{(i)} &= \partial_t \hat{v}_i\, s_i + \partial_x \hat{v}_i\, \lambda_i\, s_i \\ &= (\partial_t \hat{v}_i + \lambda_i\, \partial_x \hat{v}_i)\; s_i = 0 . \end{aligned}$$

$V_{(i)}$ $(i = 1, \ldots, m)$ is often called the *solution belonging to the i-th set of characteristics and to the given initial value* $\hat{v}_{i_0}(x)$. Obviously, the vector functions $V_{(i)}$ $(i = 1, \ldots, m)$ are linearly independent.

Remark

The fact that a sufficiently smooth solution of a nonlinear initial value problem often only exists in the neighborhood of the initial manifold, while the corresponding real-world process exists globally, means that this solution is not accepted by the physicist or engineer. The mathematician is asked to find a global solution. This forces the mathematician to create a more general definition of the solution such that the real-world situation can be described in a satisfactory way. Suitable definitions of weak solutions will be presented in Chapter 2.

Remark

As far as the scalar linear problem

$$\partial_t v + \partial_x(a\, v) = 0\ , \qquad a = \text{const}\ , \tag{1.23}$$

is concerned, v often describes a concentration, and the flux is then simply given by $f(v) = a\, v$.

Many physical processes include a further flux of the particular form

$$-\varepsilon\, \partial_x v \qquad (\varepsilon > 0)\ ,$$

proportional to the drop in concentration. The transport phenomenon is then mathematically modeled by

$$\partial_t v + \partial_x\,(a\, v - \varepsilon\, \partial_x v) = 0\ , \quad \text{i.e.,} \quad \partial_t v + a\, \partial_x v = \varepsilon\, \partial_{xx} v\ . \tag{1.24}$$

Because of the *diffusion term* on the right hand side, this equation is of parabolic type. Parabolic equations yield smoother and smoother solutions as time t increases. Therefore, shocks will be smeared out as soon as diffusion occurs.

This effect of parabolic equations can easily be demonstrated using examples like the following one. The sum

$$v(x,t) = e^{\frac{a}{2}\left(x - \frac{a}{2} t\right)} \sum_{\nu=1}^{\infty} (-1)^{\nu+1} \frac{\sin(\nu x)}{\nu} e^{-\nu^2 t}$$

converges uniformly for $t > 0$ because of the factors $e^{-\nu^2 t}$. This also holds after multiple termwise differentiations of this sum with respect to t as well as with respect to x so that v is a sufficiently smooth function for all $x \in \mathbb{R},\ t > 0$. It solves (1.24)[15] according to the classical understanding, but leads for $t = 0$ to the function

$$v_0(x) = e^{\frac{a}{2} x} \cdot w(x)$$

where $w(x)$ represents the Fourier expansion of the 2π-periodic discontinuous function

$$w(x) = \begin{cases} \frac{x}{2} & \text{for} \quad -\pi < x < \pi \\ 0 & \text{for} \quad x = \pm\pi \end{cases}\ .$$

The curve described by $w(x)$ is sometimes called a saw blade curve.

1.3
Potential Flows and (Dynamic) Buoyancy

Let us now try to investigate the forces acting on solid bodies[16] dipped into a fluid flow at a fixed position. For convenience, we restrict ourselves to stationary flows of

15) In the case of $\varepsilon = 1$

16) For example, on the wings of an aircraft

inviscid fluids. Moreover, it will be assumed that the magnitudes of the velocities are such that the density can be regarded as a constant. Thus, the flow is incompressible and the partial derivatives with respect to t occurring in (1.8) vanish. The first equation in (1.8) – if written in primitive variables, i.e., the continuity equation (1.4) – then reduces to

$$\operatorname{div} \boldsymbol{u} = 0 , \tag{1.25}$$

where[17] $\boldsymbol{u} = \left(u_1(x,y,z)\,,\; u_2(x,y,z)\,,\; u_3(x,y,z)\right)^T$ again denotes the velocity vector of the flow at the space position $\boldsymbol{x} = (x,y,z)^T$.

The second, third, and fourth equations of (1.8) formulated by means of primitive variables could be written as the vector-valued Euler equation (1.6), and they lead in the case of a stationary flow to

$$\langle \boldsymbol{u}, \nabla \rangle\ \boldsymbol{u} + \frac{1}{\varrho}\ \nabla p = \hat{\boldsymbol{k}} . \tag{1.26}$$

Let us also assume that the flow is *irrotational*; this means that the *circulation*

$$Z := \oint_C \boldsymbol{u}(\boldsymbol{x})\ \mathrm{d}\boldsymbol{x} \tag{1.27}$$

vanishes for every closed contour C within every simply connected subdomain of the flow area. Then $\boldsymbol{u}$ can be derived from a potential ϕ; i.e., there is a scalar function ϕ such that

$$\boldsymbol{u} = \nabla \phi , \tag{1.28}$$

and

$$\operatorname{curl} \boldsymbol{u} = 0 \tag{1.29}$$

holds within this area.

By the way, the vector $\operatorname{curl} \boldsymbol{u}$ is often called the *vorticity vector* or *angular velocity vector*, and a trajectory of a field of vorticity vectors is called a *vortex line*.

Because of (1.28), a flow of this type is called a *potential flow*, and ϕ is the so-called *velocity potential*.

The fifth equation in (1.8) can be omitted assuming that knowledge of the energy density E is of no interest.

From (1.29), the relation $\langle \boldsymbol{u}, \nabla \rangle\ \boldsymbol{u} = \frac{1}{2} \nabla(\|\boldsymbol{u}\|^2) - [\boldsymbol{u}, \operatorname{curl} \boldsymbol{u}]$ leads together with (1.26) to

$$\nabla \left(\frac{1}{2}\ \|\boldsymbol{u}\|^2 + \frac{1}{\varrho}\ p \right) = \hat{\boldsymbol{k}} . \tag{1.30}$$

17) For the time being, z denotes the third space variable; it will later denote the complex variable $x + i\,y$, but we have taken care to avoid any confusion.

Formula (1.30) shows that a fluid flow of the particular type under consideration, namely an approximately stationary, inviscid, incompressible, and irrotational flow can only exist if the exterior forces are the gradient of a scalar function. In other words, these forces must be conservative, i.e., there must be a potential Q with $\hat{\boldsymbol{k}} = -\nabla Q$ such that (1.30) leads to

$$\nabla \left(\frac{1}{2} \|\boldsymbol{u}\|^2 + \frac{1}{\varrho} p + Q \right) = 0 .$$

i.e., to

$$\frac{1}{2} \|\boldsymbol{u}\|^2 + \frac{1}{\varrho} p + Q = \text{const} . \tag{1.31}$$

Equation (1.31) is called the *Bernoulli equation*,[18] and is none other than the energy conservation law for this particular type of flow. p is called the *static pressure*, whereas the term $\frac{\varrho}{2} \|\boldsymbol{u}\|^2$, i.e., the kinetic energy per volume, is often called the *dynamic pressure*.

As far as incompressible flow in a circular pipe is concerned, the velocity will necessarily increase as soon as the diameter of the pipe decreases, and – because of (1.31) – this will lead to decreasing pressure within the narrow part of the pipe. This phenomenon is called the *hydrodynamic paradox*. Applications include carburetors and jet streams.

Remark

A necessary condition for the existence of irrotational flow was the conservative character of the exterior forces. Let us now assume that these forces are conservative instead, i.e., $\hat{\boldsymbol{k}} = -\nabla Q$. Moreover, let us allow the flow to be compressible, with a particular dependence $\varrho = \varrho(p)$, termed *barotropic flow*. Integration of (1.26) along a streamline from a constant point P_0 to a variable point P then leads to

$$\int_{P_0}^{P} \left\{ \frac{1}{2} \nabla(\|\boldsymbol{u}\|^2) + [\text{curl}\, \boldsymbol{u}, \boldsymbol{u}] + \nabla \Theta + \nabla Q \right\} \, \mathrm{d}s = 0 ,$$

with

$$\Theta := \int \frac{\mathrm{d}p}{\varrho(p)} ,$$

hence

$$\nabla \Theta = \frac{1}{\varrho} \nabla p ,$$

18) Daniel Bernoulli (1700–1782); St. Petersburg, Basel

and with

$$\mathrm{d}\boldsymbol{s} = \frac{1}{\|\boldsymbol{u}\|}\,\boldsymbol{u}\,\mathrm{d}s \quad (s = \text{arc length})\ .$$

Because P was arbitrary, and because of $\langle[\operatorname{curl}\boldsymbol{u},\boldsymbol{u}],\boldsymbol{u}\rangle = 0$, this result leads to the *generalized Bernoulli equation*

$$\frac{1}{2}\,\|\boldsymbol{u}\|^2 + \Theta + Q = \text{const}\ ,$$

which in the case of constant density, i.e., $\Theta = \frac{p}{\varrho}$, seems to coincide completely with (1.31). However, it should be noted that the constant on the right hand side can now change from streamline to streamline. Additionally, we find

$$\begin{aligned}
\frac{\mathrm{d}}{\mathrm{d}t}Z &= \frac{\mathrm{d}}{\mathrm{d}t}\oint_C \boldsymbol{u}(\boldsymbol{x})\,\mathrm{d}\boldsymbol{x} = \oint_C \frac{\mathrm{d}}{\mathrm{d}t}\left\langle \boldsymbol{u}, \frac{\mathrm{d}\boldsymbol{x}}{\mathrm{d}s}\right\rangle \mathrm{d}s \\
&= \oint_C \frac{\mathrm{d}\boldsymbol{u}}{\mathrm{d}t}\,\mathrm{d}\boldsymbol{x} + \oint_C \boldsymbol{u}\,\mathrm{d}\boldsymbol{u} \\
&= \oint_C \frac{\mathrm{d}\boldsymbol{u}}{\mathrm{d}t}\,\mathrm{d}\boldsymbol{x} = \oint_C \{\langle \boldsymbol{u},\nabla\rangle\,\boldsymbol{u}\}\,\mathrm{d}\boldsymbol{x} \\
&= -\oint_C \left\{\frac{1}{\varrho}\nabla p + \nabla Q\right\}\mathrm{d}\boldsymbol{x} \\
&= -\oint_C \nabla\left(\Theta + Q\right)\,\mathrm{d}\boldsymbol{x} = 0\ .
\end{aligned}$$

This is *Kelvin's theorem*,[19] which says that the circulation along a closed curve in an inviscid barotropic flow does not change over time.

If an irrotational flow with an arbitrary type of incompressibility is considered, (1.28) leads together with (1.25) to

$$\Delta\phi = 0 \tag{1.32}$$

whereas, for a compressible fluid in the case of stationary flow, the continuity equation (1.4) is

$$\operatorname{div}(\varrho\boldsymbol{u}) = 0\ .$$

Let us additionally assume the flow to be barotropic. Then, because of

$$\operatorname{div}(\varrho\boldsymbol{u}) = \varrho\operatorname{div}\boldsymbol{u} + \frac{\mathrm{d}\varrho}{\mathrm{d}p}\,\langle\nabla p,\boldsymbol{u}\rangle$$

19) Lord Kelvin of Largs (1824–1907); Glasgow

and because of

$$\hat{c} = \sqrt{\frac{\mathrm{d}p}{\mathrm{d}\varrho}} \quad \text{(cf. (1.67))} ,$$

(1.26) leads to

$$\left(1 - \left(\frac{u_1}{\hat{c}}\right)^2\right) \partial_{xx}\phi + \left(1 - \left(\frac{u_2}{\hat{c}}\right)^2\right) \partial_{yy}\phi + \left(1 - \left(\frac{u_3}{\hat{c}}\right)^2\right) \partial_{zz}\phi$$
$$- 2\left(\frac{u_1\, u_2}{\hat{c}^2} \cdot \partial_y u_1 + \frac{u_2\, u_3}{\hat{c}^2} \cdot \partial_z u_2 + \frac{u_3\, u_1}{\hat{c}^2} \cdot \partial_x u_3\right) = 0 , \tag{1.33}$$

as far as the exterior forces vanish. In this situation, (1.33) generalizes (1.32).

Obviously, (1.33) is a quasilinear partial differential equation of second order for ϕ, which is certainly elliptic if

$$M := \frac{\|\boldsymbol{u}\|}{\hat{c}} < 1 , \tag{1.34}$$

i.e., in areas of *subsonic flow*.

Definition

M is called the *Mach number.*[20]

In particular, if we consider a constant flow in the *x*-direction which is only disturbed in the neighborhood of a slim airfoil[21] with a small angle of attack, the products of the values u_i $(i = 2, 3)$ with each other and with u_1 can be neglected when compared with 1. This leads to $\|\boldsymbol{u}\|^2 = u_1^2$ and to a shortened version of (1.33), namely to

$$\left(1 - \left(\frac{u_1}{\hat{c}}\right)^2\right) \partial_{xx}\phi + \partial_{yy}\phi + \partial_{zz}\phi = 0 . \tag{1.35}$$

This equation is hyperbolic in areas of *supersonic flow* $(M > 1)$, and the full equation (1.33) is also hyperbolic in this case.

Let us extend our idealizations by assuming that the flow under consideration is a two-dimensional *plane flow*. This means that one of the components of the velocity vector $\boldsymbol{u}$ in a rectangular coordinate system, e.g., the component in the *z*-direction, vanishes for all $(x, y) \in \mathbb{R}^2$ and for all $t \geq 0$:

$$u_3 = 0 . \tag{1.36}$$

In the case of a two-dimensional supersonic flow along a slim airfoil for which (1.35) holds, *Mach's angle* β, determined from

$$|\sin\beta| = \frac{1}{M} < 1 ,$$

20) Ernst Mach (1838–1916); Graz, Praha, Vienna

21) A cross-section of a wing or another rigid body in a plane parallel to the direction of the flow

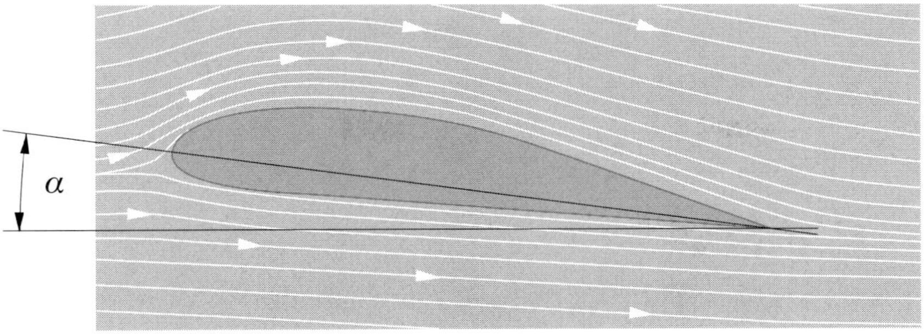

Fig. 1.3 Flow around a wing if the angle of attack is small.

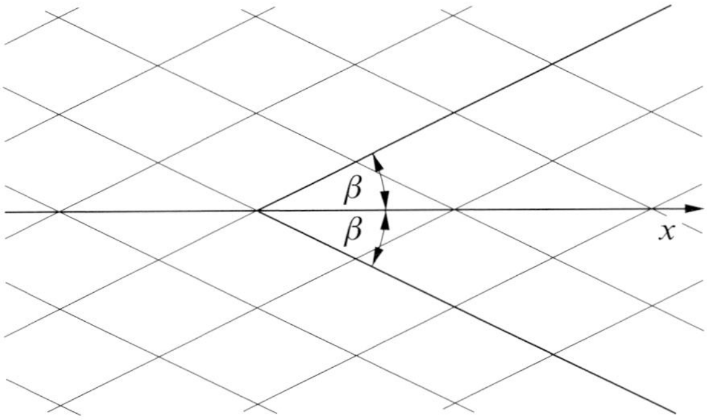

Fig. 1.4 Mach's net in the case of a linearized supersonic flow (M = const > 1).

describes the angle between the characteristics of the wave equation

$$(1 - M^2)\partial_{xx}\phi + \partial_{yy}\phi = 0$$

and the flow direction given by the direction of the x-axis. The set of all of these characteristics is called *Mach's net*. This net plays an important role when so-called *methods of characteristics* are used in order to establish efficient numerical procedures.

The two-dimensional case of rotational symmetry, e.g., flow along a projectile, leads correspondingly to *Mach's cone*.

In the two-dimensional plane situation, (1.29) becomes

$$\operatorname{curl} \boldsymbol{u} = \begin{pmatrix} -\partial_z u_2 \\ \partial_z u_1 \\ \partial_x u_2 - \partial_y u_1 \end{pmatrix} = 0 \, .$$

Thus, u_1 and u_2 are independent of the third spatial variable z:

$$\boldsymbol{u} = \begin{pmatrix} u_1(x, y) \\ u_2(x, y) \\ 0 \end{pmatrix}$$

with

$$\partial_x u_2 - \partial_y u_1 = 0 \,. \tag{1.37}$$

After introducing the vector

$$\boldsymbol{v} := \begin{pmatrix} -u_2(x,y) \\ u_1(x,y) \\ 0 \end{pmatrix} \tag{1.38}$$

and using the continuity equation (1.25), we obtain

$$\operatorname{curl} \boldsymbol{v} = \begin{pmatrix} 0 \\ 0 \\ \partial_x u_1 + \partial_y u_2 \end{pmatrix} = \begin{pmatrix} 0 \\ 0 \\ \operatorname{div} u \end{pmatrix} = 0 \,,$$

such that $\boldsymbol{v}$ can also be derived from the potential in simply connected parts of the fluid area. In other words, there is a scalar function $\psi = \psi(x, y)$ with

$$-u_2 = \partial_x \psi \,, \quad u_1 = \partial_y \psi \,. \tag{1.39}$$

ψ is called the *stream function.* (1.28) leads to

$$\partial_x \phi = \partial_y \psi \,, \quad \partial_y \phi = -\partial_x \psi \,. \tag{1.40}$$

Obviously, (1.40) can be interpreted as the system of *Cauchy–Riemann equations* of the complex function

$$\Omega(z) := \phi(x,y) + i\,\psi(x,y) \tag{1.41}$$

depending on the complex variable $z = x + i\,y$.

Ω is called the *complex velocity potential* of the plane potential flow under consideration.

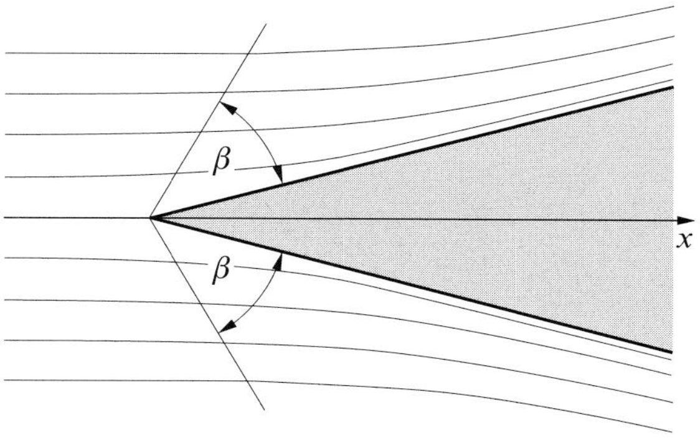

Fig. 1.5 Linearized supersonic flow around a slim rotatory cone.

We are going to assume that the first partial derivatives of the functions ϕ and ψ are continuous such that Ω is found to be a holomorphic function, and we intend to study the forces acting on rigid bodies when dipped into such a fluid flow.

Because we reduced reality to a plane flow, the rigid body is assumed to be very long with respect to the direction of the third spatial variable; more precisely, it must be of infinite length from the point of view of mathematics. Therefore, the flow around the contour of the cross-section of the body within the (x, y)-plane, e.g., around the contour of an airfoil of a long wing, is of interest.

We stated that the circulation $Z = \oint_C \boldsymbol{u}\, d\boldsymbol{x}$ vanishes as long as the closed contour C is the boundary of a simply connected domain within the fluid area in the case of a potential flow. Because of our plane model, C represents a simple closed contour in the (x, y)-plane, i.e., in $\mathbb{R}^2$.

Now, consider the situation where the boundary Γ of a rigid body is dipped into the fluid, e.g., a wing. If its airfoil is part of the area surrounded by C, this interior domain of C is no longer a simply connected domain of the fluid area. Hence, the circulation Z around the airfoil does not necessarily vanish but is found to fulfill

$$\oint_C \boldsymbol{u}\, d\boldsymbol{x} = \oint_\Gamma \boldsymbol{u}\, d\boldsymbol{x}\,.$$

In order to proof this relation, we take into account that Γ and C can, in a first step, be connected by two auxiliary lines in such a way that two simply connected domains G_1 and G_2, with contours C_1 and C_2, respectively, will occur. Because of $\operatorname{curl} \boldsymbol{u} = 0$ in G_1 as well as in G_2, we see in a second step that

$$\oint_{C_1} \boldsymbol{u}\, d\boldsymbol{x} = 0 \quad \text{as well as} \quad \oint_{C_2} \boldsymbol{u}\, d\boldsymbol{x} = 0$$

holds. This leads to

$$\oint_{C_1} \boldsymbol{u}\, d\boldsymbol{x} + \oint_{C_2} \boldsymbol{u}\, d\boldsymbol{x} = 0$$

(cf. Fig. 1.6).

We realize in a third step that the integrations back and forth along each of the auxiliary lines extinguish each other such that

$$\oint_{C_1} \boldsymbol{u}\, d\boldsymbol{x} + \oint_{C_2} \boldsymbol{u}\, d\boldsymbol{x} = \oint_C \boldsymbol{u}\, d\boldsymbol{x} + \oint_{-\Gamma} \boldsymbol{u}\, d\boldsymbol{x}$$

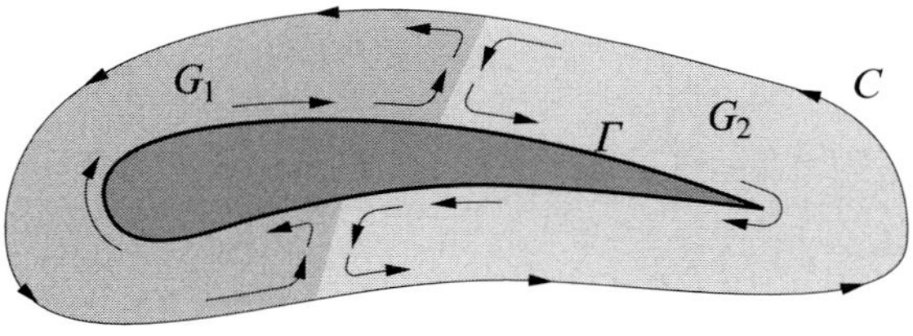

Fig. 1.6 Auxiliary step in the computation of the buoyancy generated by plane potential flows.

results. Here, $-\Gamma$ denotes the contour of the rigid body along the negative direction, and this ends the proof.

Remark

It should be noted that the proof given here is simplified because Γ not only passes through the interior of the domain occupied by the fluid but is also part of its boundary. As a matter of fact, one must first, roughly speaking, investigate the case where Γ is replaced by a line Γ_ε of distance ε from Γ that passes only through the fluid, and in a next step one must study the limit situation $\varepsilon \longrightarrow 0$.

Let Γ now be parametrized by $\boldsymbol{r} = \boldsymbol{r}(\tau)$, $0 \leq \tau \leq T$. Then

$$Z = \int_0^T \langle \boldsymbol{u}(\boldsymbol{r}(\tau)), \dot{\boldsymbol{r}}(\tau)\rangle \, \mathrm{d}\tau = \int_0^T \{u_1\, \dot{x} + u_2\, \dot{y}\} \, \mathrm{d}\tau \tag{1.42}$$

holds, and $\dot{\boldsymbol{r}} = \begin{pmatrix} \dot{x}(\tau) \\ \dot{y}(\tau) \end{pmatrix}$ is a vector tangential to the curve Γ at the point $(x(\tau), y(\tau))$ such that

$$\boldsymbol{n} := \frac{1}{\sqrt{\dot{x}^2 + \dot{y}^2}} \begin{pmatrix} \dot{y} \\ -\dot{x} \end{pmatrix} \tag{1.43}$$

is a normal unit vector of Γ at this point.

According to (1.13), we assume that the velocity of the flow at the surface of the rigid body is tangential to this surface:

$$\langle \boldsymbol{u}, \boldsymbol{n}\rangle = 0 \quad \text{along} \quad \Gamma \, ,$$

i.e., $u_1\, \dot{y} = u_2\, \dot{x}$. From this, and with (1.42), the circulation becomes

$$Z = \int_0^T (u_1 - i\, u_2)\, (\dot{x} + i\, \dot{y}) \, \mathrm{d}\tau \, ,$$

i.e.,

$$Z = \int_\Gamma w(z) \, \mathrm{d}z \, , \tag{1.44}$$

with

$$w(z) := u_1(x, y) - i\, u_2(x, y) \, . \tag{1.45}$$

$w(z)$ is a holomorphic function in the whole domain outside the airfoil because (1.39) yields

$$\partial_x u_1 = \partial_{xy}\psi = \partial_{yx}\psi = \partial_y(-u_2) \, ,$$

and because (1.28) leads to

$$\partial_y u_1 = \partial_{xy}\phi = \partial_{yx}\phi = -\partial_x(-u_2) \, ,$$

the Cauchy–Riemann equations for the function $w(z)$ are fulfilled.

Hence, also with respect to the computation of the circulation, integration along C instead of Γ is permitted:

$$Z = \int_C w(z)\,dz\,. \tag{1.46}$$

Here, we choose C in such a way that it lies in the annulus between two concentric circles, where circle K_r of radius r surrounds the airfoil and K_R is a circle with a sufficiently great arbitrary radius $R > r$. Without any loss of generality, we assume the center of each circle to be the origin of the former (x, y)-plane, which now becomes the complex z-plane.

$w(z)$ can be represented in the annulus (and therefore for every $z \in C$ in particular) by a Laurent series[22] around the center $z_0 = 0$ (cf. Fig. 1.7):

$$w(z) = \sum_{\nu=-\infty}^{+\infty} a_\nu\, z^\nu\,.$$

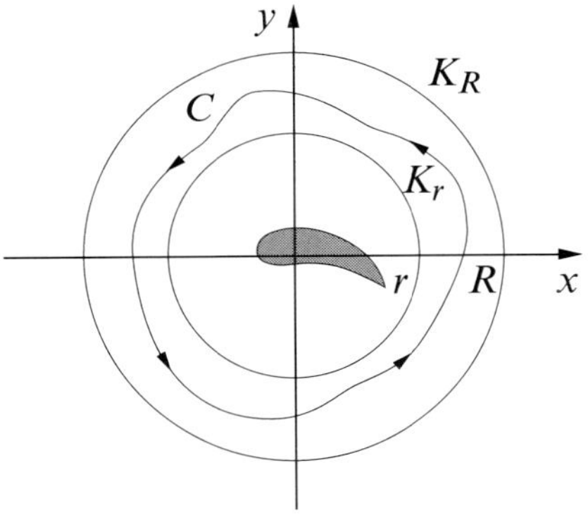

Fig. 1.7 Annulus around an airfoil.

We know from experience that the fluid flow is often only influenced by the rigid body within the neighborhood (which has a certain size) of the contour; the flow around a ship that crosses a calm lake at a constant velocity provides an example of this. From this point of view, we are going to assume that the velocity $\boldsymbol{u} = \boldsymbol{u}(x, y)$ is constant for $|z| \to \infty$:

$$\lim_{|z|\to\infty} \boldsymbol{u}(x, y) = \begin{pmatrix} u_{1,\infty} \\ u_{2,\infty} \end{pmatrix} \tag{1.47}$$

with constant components $u_{1,\infty}$ and $u_{2,\infty}$ such that

$$\lim_{|z|\to\infty} w(z) = u_{1,\infty} - i\,u_{2,\infty}\,.$$

22) Pierre Alphonse Laurent (1813–1854); Le Havre

Because the coefficients a_ν ($\nu = 0, \pm 1, \pm 2, \ldots$) are constant numbers which do not depend on R, and because R is allowed to tend to infinity, we find for $|z| \to \infty$ that all the a_ν with positive indices must vanish.

Hence, the result

$$\lim_{|z|\to\infty} w(z) = a_0 = u_{1,\infty} - i\, u_{2,\infty} \tag{1.48}$$

follows.

On the other hand, Cauchy's formula

$$a_{-1} = \frac{1}{2\pi i} \oint_C w(\zeta)\, \mathrm{d}\zeta$$

leads to

$$Z = 2\pi i\, a_{-1} \ . \tag{1.49}$$

The (two-dimensional) force $\boldsymbol{K}$ to be calculated, which is caused by the flow and acts on the rigid body, is found to be:

$$\boldsymbol{K} = \begin{pmatrix} K_1 \\ K_2 \end{pmatrix} = -\int_0^L p\, \boldsymbol{n}\, \mathrm{d}s \quad \left([\boldsymbol{K}] = \frac{[\text{force}]}{[\text{length}]} \right) , \tag{1.50}$$

where $\boldsymbol{n}$ is the unit normal vector from (1.43), p denotes the pressure along the surface of the body caused by the fluid, s measures the arc length along Γ beginning at an arbitrary point on it, and where L is the total length of Γ. Let us study this force separately from the other exterior forces acting on the body; i.e., let us assume that the sum of these other forces vanishes. This leads to a constant potential Q in Bernoulli's equation (1.31) and therefore to a constant *total pressure*

$$p_0 := \frac{\varrho}{2}\, \|\boldsymbol{u}\|^2 + p \ .$$

Because Γ is a closed contour, $\int_0^L \boldsymbol{n}\, \mathrm{d}s = 0.$[23)]
Therefore,

$$\boldsymbol{K} = -\int_0^L \left\{ p_0 - \frac{\varrho}{2}\, \|\boldsymbol{u}\|^2 \right\} \boldsymbol{n}\, \mathrm{d}s = \frac{\varrho}{2} \int_0^L \|\boldsymbol{u}\|^2\, \boldsymbol{n}\, \mathrm{d}s \ ,$$

23) $\dfrac{\mathrm{d}s}{\mathrm{d}\tau} = \sqrt{\dot{x}^2 + \dot{y}^2}$, hence

$$\int_0^L \boldsymbol{n}\, \mathrm{d}s = \int_0^T \frac{1}{\sqrt{\dot{x}^2 + \dot{y}^2}} \begin{pmatrix} \dot{y} \\ -\dot{x} \end{pmatrix} \frac{\mathrm{d}s}{\mathrm{d}\tau}\, \mathrm{d}\tau = \begin{pmatrix} y(T) - y(0) \\ -x(T) + x(0) \end{pmatrix} = \begin{pmatrix} 0 \\ 0 \end{pmatrix}$$

i.e.,

$$K_1 = \frac{\varrho}{2} \int_0^T \frac{1}{\sqrt{\dot{x}^2 + \dot{y}^2}} \|\boldsymbol{u}\|^2 \dot{y} \frac{\mathrm{d}s}{\mathrm{d}\tau} \, \mathrm{d}\tau ,$$

$$K_2 = -\frac{\varrho}{2} \int_0^T \frac{1}{\sqrt{\dot{x}^2 + \dot{y}^2}} \|\boldsymbol{u}\|^2 \dot{x} \frac{ds}{d\tau} \, d\tau .$$

The temporary introduction of the complex number

$$k = K_2 + i\, K_1 \tag{1.51}$$

yields

$$k = -\frac{\varrho}{2} \int_0^T \frac{1}{\sqrt{\dot{x}^2 + \dot{y}^2}} \|\boldsymbol{u}\|^2 (\dot{x} - i\,\dot{y}) \frac{\mathrm{d}s}{\mathrm{d}\tau} \, \mathrm{d}\tau = -\frac{\varrho}{2} \int_0^T \|\boldsymbol{u}\|^2 (\dot{x} - i\,\dot{y}) \, \mathrm{d}\tau .$$

The complex number within the parentheses on the right hand side of

$$\begin{aligned} \|\boldsymbol{u}\|^2 (\dot{x} - i\,\dot{y}) &= \left(u_1^2 + u_2^2\right) (\dot{x} - i\,\dot{y}) = (u_1 - i\,u_2)(u_1 + i\,u_2)(\dot{x} - i\,\dot{y}) \\ &= w(z) \, (u_1\,\dot{x} + u_2\,\dot{y} + i\,u_2\,\dot{x} - i\,u_1\,\dot{y}) \end{aligned}$$

is not really complex because the imaginary part vanishes (cf. (1.13)). It can therefore be replaced by its conjugate complex number:

$$\begin{aligned} \|\boldsymbol{u}\|^2 (\dot{x} - i\,\dot{y}) &= w(z) \, (u_1\,\dot{x} + u_2\,\dot{y} - i\,u_2\,\dot{x} + i\,u_1\,\dot{y}) \\ &= w(z) \, (u_1 - i\,u_2) \, (\dot{x} + i\,\dot{y}) = w^2(z)\, \dot{z} . \end{aligned}$$

This leads to

$$k = -\frac{\varrho}{2} \int_0^T w^2(z)\, \dot{z} \, \mathrm{d}\tau = -\frac{\varrho}{2} \int_\Gamma w^2(z) \, \mathrm{d}z .$$

Because w is holomorphic, w^2 is also a holomorphic function. Hence,

$$k = -\frac{\varrho}{2} \int_C w^2(z) \, \mathrm{d}z . \tag{1.52}$$

However, within the annulus we have

$$\begin{aligned} w^2(z) &= \left(a_0 + \frac{a_{-1}}{z} + \frac{a_{-2}}{z^2} + \ldots\right) \left(a_0 + \frac{a_{-1}}{z} + \frac{a_{-2}}{z^2} + \ldots\right) \\ &= a_0^2 + 2\,\frac{a_0\,a_{-1}}{z} + \frac{A_{-2}}{z^2} + \frac{A_{-3}}{z^3} + \ldots \end{aligned}$$

with certain coefficients A_ν ($\nu = -2, -3, \ldots$).

Cauchy's residuum formula and (1.52) therefore lead to

$$k = -\frac{\varrho}{2}\, 2a_0\, a_{-1} \cdot 2\pi i\,.$$

Thus, (1.48) and (1.49) yield

$$k = -\varrho\, (u_{1,\infty} - i\, u_{2,\infty})\, Z\,,$$

and because the circulation Z is real (cf. (1.27)), comparison of the real and imaginary parts of (1.51) results in

$$\begin{aligned} K_1 &= \varrho\, u_{2,\infty}\, Z \\ K_2 &= -\varrho\, u_{1,\infty}\, Z\,. \end{aligned} \tag{1.53}$$

In particular, if $u_{2,\infty} = 0$ but $u_{1,\infty} \neq 0$, i.e., if the undisturbed flow is parallel to the x-axis, and if $Z \neq 0$, we obtain a *lift* $K_2 \neq 0$, i.e., there is a force acting on the rigid body perpendicular to the direction of the flow.

By using an appropriate wing construction, the airfoils can yield $Z < 0$, so that an aircraft can lift its own weight, a hydrofoil can rise out of the water, etc. Of course, Archimedes' static buoyancy, as given by (1.4), must also be taken into account.

Remark

The formulae (1.53) are called *Kutta–Zhukovsky buoyancy formulae.*[24)]

Remark

Whereas the first equation in (1.53) describes the buoyancy at least qualitatively in a correct manner, the result $K_1 = 0$ contradicts reality. Also, in the case of an incompressible irrotational stationary flow (as more or less realized by calmly flowing streams), the flow will apply some force to the rigid body (e.g., a bridge pier) parallel to the direction of flow. Of course, this contradiction results from one of our idealizations: the assumption of an inviscid and therefore frictionless fluid.

In order to understand what really happens parallel to the flow, we must reduce the amount of idealization. This will later be achieved by using the so-called Navier–Stokes equations rather than the Euler equations, and the so-called *no-slip condition* given below rather than (1.13).

$$\boldsymbol{u} = 0 \quad \text{along} \quad \Gamma\,. \tag{1.54}$$

Equation (1.54) expresses the idea that moving viscous fluids leave a monomolecular stationary layer on impermeable walls of solid bodies because of adhesion. Thus, friction along such surfaces does not mean friction between the fluid and the solid material, but rather friction between fluid particles.

Of course, there are also other boundary conditions where *partial slip* occurs on the surface, e.g., in rarefied gas flow, with porous walls, etc. The tangential component of the flow is then proportional to the local shear stress.

24) Martin Wilhelm Kutta (1867–1927); Stuttgart; Nikolai Jegorowitsch Zhukovsky (1847–1921); Moscow

If (1.33) is reduced to the case of an irrotational, stationary, plane flow (i.e., $u_3 = 0$, $\partial_y u_1 = \partial_x u_2$), if there are no exterior forces, and if (u_1, u_2) is for convenience replaced with (u, v), we get:

$$\left(1 - \left(\frac{u}{\hat{c}}\right)^2\right) \partial_x u + \left(1 - \left(\frac{v}{\hat{c}}\right)^2\right) \partial_y v - \frac{u\, v}{\hat{c}^2} \cdot (\partial_y u + \partial_x v) = 0 \,. \tag{1.55}$$

The uv-plane is called the *hodograph plane*.

We assume the equations

$$u = u(x, y)$$
$$v = v(x, y)$$

to be invertible such that x and y can be expressed by u and v, i.e.,

$$D := \begin{vmatrix} \partial_x u & \partial_y u \\ \partial_x v & \partial_y v \end{vmatrix} \neq 0 \,.$$

In other words, it will be assumed that the vectors $\partial_x \boldsymbol{u}$ and $\partial_y \boldsymbol{u}$ are linearly independent.

This leads immediately to

$$\partial_x u = D\, \partial_v y \,, \quad \partial_y u = -D\, \partial_v x \,, \quad \partial_x v = -D\, \partial_u y \,, \quad \partial_y v = D\, \partial_u x \,.$$

The nonlinear equation (1.55) together with the equation $\partial_y u = \partial_x v$ for the functions $u(x, y)$ and $v(x, y)$ of the irrotational flow can therefore – after division by D – be transformed into the linear equations

$$\left(1 - \frac{v^2}{\hat{c}^2}\right) \partial_u x + \left(1 - \frac{u^2}{\hat{c}^2}\right) \partial_v y + \frac{uv}{\hat{c}^2} \left(\partial_v x + \partial_u y\right) = 0$$
$$\partial_v x - \partial_u y = 0 \tag{1.56}$$

for the functions $x(u, v)$ and $y(u, v)$.

The transition from the original equations to the linear equations (1.56) is called the *method of hodographs* and corresponds to the *Legendre transformation* in the theory of partial differential equations.

In simply connected domains, $\boldsymbol{x}(\boldsymbol{u})$ can be derived from a potential Θ (i.e., $x = \Theta_u$, $y = \Theta_v$), because of

$$\operatorname{curl} \boldsymbol{x}(\boldsymbol{u}) = (0, 0, \partial_u y - \partial_v x)^T = 0 \,.$$

Equation (1.56) therefore yields

$$\left(1 - \frac{v^2}{\hat{c}^2}\right) \partial_{uu} \Theta + \left(1 - \frac{u^2}{\hat{c}^2}\right) \partial_{vv} \Theta + 2 \frac{uv}{\hat{c}^2} \partial_{uv} \Theta = 0 \,. \tag{1.57}$$

If polar coordinates ($w = \|\boldsymbol{u}\|$, α) are used in the hodograph plane, i.e.,

$$u = w \cos \alpha \,, \quad v = w \sin \alpha \,,$$

$$\partial_u = \cos \alpha\, \partial_w - \frac{1}{w} \sin \alpha\, \partial_\alpha \,,$$

$$\partial_v = \sin \alpha\, \partial_w + \frac{1}{w} \cos \alpha\, \partial_\alpha \,,$$

Equation (1.57) becomes the so-called *hodograph equation*:

$$\partial_{ww}\Theta + \frac{1}{w^2}\left(1 - \frac{w^2}{\hat{c}^2}\right)\partial_{\alpha\alpha}\Theta + \frac{1}{w}\left(1 - \frac{w^2}{\hat{c}^2}\right)\partial_w\Theta = 0 \qquad (1.58)$$

which does not explicitly contain α.

If we try to solve (1.58) by the ansatz

$$\Theta(w, \alpha) = g(w)\sin(m\alpha) \quad \text{or} \quad \Theta(w, \alpha) = g(w)\cos(m\alpha)\ , \qquad (m \in \mathbb{R})\ , \qquad (1.59)$$

we obtain for the unknown function $g(w)$ the ordinary differential equation

$$g''(w) + \frac{1}{w}\left(1 - \frac{w^2}{\hat{c}^2}\right)g'(w) - \frac{m^2}{w^2}\left(1 - \frac{w^2}{\hat{c}^2}\right)g(w) = 0\ . \qquad (1.60)$$

Solutions of the hodograph equation of type (1.59) are called *Chapligin solutions.*[25)]

Let $g_m(w)$ be a solution of (1.60) that belongs to a particular m and let $\Theta_m(w, \alpha)$ be the solution of (1.59) that corresponds to this solution.

Examples:

$m = 0$ leads to

$$\frac{g_0''}{g_0'} = -\frac{1}{w} + \frac{w}{\hat{c}^2}\ ,$$

hence

$$g_0' = \frac{c_1^{(0)}}{w}e^{\frac{1}{2}\frac{w^2}{\hat{c}^2}}\ .$$

The power series of $e^{\frac{1}{2}\frac{w^2}{\hat{c}^2}}$ converges uniformly for all values of w. Integration can therefore be performed term-by-term, yielding

$$g_0(w) = c_1^{(0)}\left\{\ln w + \sum_{\nu=1}^{\infty}\frac{\left(\frac{w}{\hat{c}}\right)^{2\nu}}{2^{\nu+1}\nu\nu!}\right\} + c_2^{(0)}\ .$$

Here, $c_1^{(0)}$ and $c_2^{(0)}$ are arbitrary constants.

Analogously for $m = 1$:

$$g_1(w) = c_1^{(1)}w + c_2^{(1)}\left\{\frac{1}{w^2} + \frac{w\ln w}{2\hat{c}^2} + \sum_{\nu=0}^{\infty}\frac{w^{2(\nu+1)}}{2^{\nu+2}\hat{c}^{2(\nu+2)}}\right\}\ .$$

Because (1.58) is linear and homogeneous, all linear combinations of the particular Chapligin solutions also solve Eq. (1.58). The coefficients of the expansion as well as the constants $c_1^{(m)}$ and $c_2^{(m)}$ must be chosen in such a way that the expansion fits the given situation, at least approximately.

25) C.A. Chapligin: Sci. Ann. Univ. Moscow. Math. Phys. **21** (1904) 1–121

1.4
Motionless Fluids and Sound Propagation

Obviously, because viscosity only plays a role in moving fluids, the results of this section are also valid for real fluids.

In a first step, let us consider the case for constant density ϱ, which is approximately realized in liquids.

In this situation, (1.26) leads for motionless fluids (i.e., for $\boldsymbol{u} = 0$) to the so-called *hydrostatic equation*

$$\hat{\boldsymbol{k}} = \nabla \left(\frac{p}{\varrho} \right) ,$$

and (1.31) becomes

$$Q + \frac{p}{\varrho} = \text{const} .$$

Let us assume that we do not yet know how the free surface of a motionless liquid behaves if only the force of gravity and a constant exterior (e.g., atmospheric) pressure p_0 affects this liquid. Hence, the force per mass unit is given by

$$\hat{\boldsymbol{k}} = (0,\ 0,\ -g)^T \quad (g = \text{acceleration due to gravity}) ,$$

such that $Q = gz + \text{const}$ (i.e., $\varrho g z + p = \text{const}$). In particular, if (x, y, z_0) is a point on the surface, $\varrho g z + p = \varrho g z_0 + p_0$ or

$$\varrho g (z - z_0) = -(p - p_0) = -\hat{p} \tag{1.61}$$

holds. Here, $\hat{p}$ is the overpressure inside the liquid compared with the exterior pressure.

Because of

$$z_0 = \frac{\text{const} - p_0}{\varrho g} ,$$

z_0 is constant, and thus independent of (x, y). In other words, the surface of the liquid is a plane, or more precisely it is parallel to the Earth's surface. If $h = z_0 - z$ is the height of an arbitrarily shaped liquid column, (1.61) gives

$$\varrho g h q = \hat{p} q ,$$

where q is the base area of the column.

The force affecting this base is given by the right hand side of the equation and does not depend on the form of the column, whereas the left hand side gives the weight of a *cylindrical* column of the fluid with the same base area and the same height.

This phenomenon is called the *hydrostatic paradox*.

We are now going to dip a solid body of volume V and surface F into a stationary liquid.

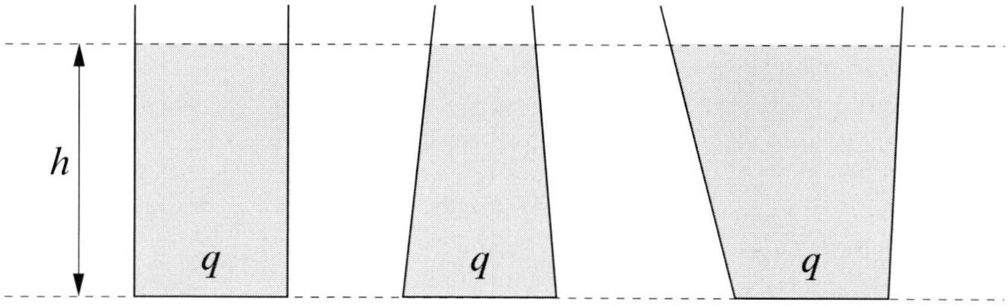

Fig. 1.8 Hydrostatic paradox.

Obviously, an overall force $\boldsymbol{K}$ affects the body, where

$$\boldsymbol{K} = (K_1, K_2, K_3)^T = -\int_F p\,\boldsymbol{n}\,\mathrm{d}o + \boldsymbol{G}\,.$$

Here, p is the interior pressure of the liquid, $\boldsymbol{G}$ is the weight of the body, and $\boldsymbol{n}$ denotes the outward-directed normal unit vector at the points on the surface. Because of $\boldsymbol{G} = (0\,,\ 0\,,\ -G)^T$, (1.61) yields

$$K_1 = \varrho\, g \int_F \left(z - z_0 - \frac{p_0}{\varrho\, g}\right) n_1\,\mathrm{d}o = \varrho\, g \int_F \langle \boldsymbol{a}, \boldsymbol{n}\rangle\,\mathrm{d}o$$

with $\boldsymbol{a} := (z - z_0 - \dfrac{p_0}{\varrho\, g}, 0, 0)^T$. The divergence theorem therefore leads to

$$K_1 = \varrho\, g \int_V \operatorname{div} \boldsymbol{a}\,\mathrm{d}V = \varrho\, g \int_V \frac{\partial(z - z_0)}{\partial x}\,\mathrm{d}V = 0\,.$$

Analogously, $K_2 = 0$. K_3 is found to be given by

$$K_3 = \varrho\, g \int_V \frac{\partial(z - z_0)}{\partial z}\,\mathrm{d}V - G = \varrho\, g\, V - G\,,$$

so that

$$\boldsymbol{K} = (0, 0, \varrho\, g\, V - G)^T\,. \tag{1.62}$$

ϱ is the density of the liquid (!), so $\varrho\, g\, V$ is the weight of the particular part of the liquid which is displaced by the solid body. Hence, the body is affected by a force directed against the direction of action of its weight, and this (static) buoyancy equals the weight of the displaced quantity of the liquid (*Archimedes' principle*).[26)]

In order to study the propagation of sound in a fluid, we assume that the fluid does not move from a macroscopic point of view. Moreover, we will only consider the sound effects resulting from very small changes in the density and pressure

26) Archimedes (about 220 BC); Syracuse

within the fluid. Therefore, we no longer assume the density to be constant, but we assume only small variations in it.

Let $\hat{\varrho}$ and $\hat{p}$ be the averages of the density and the pressure, respectively. Only these averages are expected to be constant quantities.

The density disturbances will be expressed by

$$\varrho = \hat{\varrho}\ \left(1 + \sigma(\boldsymbol{x}, t)\right)$$

where $|\sigma| \ll 1$ and the spatial derivatives of σ are small.[27)]

The continuity equation (1.4) then becomes

$$\hat{\varrho}\,\sigma_t + \hat{\varrho}\ \mathrm{div}\left((1 + \sigma)\boldsymbol{u}\right) = 0\ ,$$

and so, after dividing by $\hat{\varrho}$ and taking the assumptions for σ into account,

$$\sigma_t + \mathrm{div}\,\boldsymbol{u} = 0\ . \tag{1.63}$$

Experiments show that the propagation of sound waves occurs more or less adiabatically, i.e., without any gain or loss of heat. The equation of state to be taken into account is therefore

$$\varrho^{-\gamma}\, p = \mathrm{const}^{28)} \tag{1.64}$$

such that

$$\frac{p}{\hat{p}} = \left(\frac{\varrho}{\hat{\varrho}}\right)^{\gamma} = (1 + \sigma)^{\gamma}\ .$$

Because σ is small,

$$(1 + \sigma)^{\gamma} \approx 1 + \gamma\,\sigma\ .$$

Hence

$$p = \hat{p}(1 + \gamma\,\sigma)\ ,$$

so that

$$\frac{1}{\varrho}\,\nabla p = \frac{\hat{p}\,\gamma}{(1 + \sigma)\,\hat{\varrho}}\,\nabla\sigma\ . \tag{1.65}$$

In our model of small disturbances, the velocity of the fluid particles and its spatial derivatives are so small that higher-order terms of these quantities can be neglected

27) Our results do not necessarily hold in the case of the propagation of *large* variations in the pressure or density, such as can occur in the case of detonations.

28) γ from (1.9)

compared with first-order terms. In other words, the convection term in (1.6) can be neglected. Also, exterior forces do not play a role in our scenario.

If (1.65) is then put into (1.6), we find

$$\boldsymbol{u}_t = -\frac{\gamma\,\hat{p}}{\hat{\varrho}}\,\nabla\sigma\,,$$

where $1+\sigma$ was approximated by 1.

Forming the divergence of this term, and then changing the sequence of the time derivative and the spatial derivatives, we end up with

$$(\operatorname{div}\boldsymbol{u})_t = -\frac{\gamma\,\hat{p}}{\hat{\varrho}}\,\Delta\sigma\,.$$

This result can be compared with (1.63) when differentiated with respect to t. This comparisons yields

$$\sigma_{tt} = \frac{\gamma\,\hat{p}}{\hat{\varrho}}\,\Delta\,\sigma\,,$$

a wave equation for σ. It shows that the sound waves propagate within the fluid with the velocity

$$\hat{c} = \sqrt{\frac{\gamma\,\hat{p}}{\hat{\varrho}}}\,. \qquad (1.66)$$

Definition

$\hat{c}$ is called the *local speed of sound.*

Remark

Because of (1.64), $\hat{c}$ can also be represented by

$$\hat{c} = \sqrt{\frac{\mathrm{d}p}{\mathrm{d}\varrho}}\,. \qquad (1.67)$$

2
Weak Solutions of Conservation Laws

2.1
Generalization of What Will Be Called a Solution

As already announced, in this chapter we are going to discuss the introduction of a suitable globalization of the definition of a conservation law solution. We already know that such a solution cannot be expected to be a smooth function, because we noted that discontinuities can even arise from smooth initial states. Moreover, discontinuities that arise at an instant t_0 can move with respect to space and time, as will be seen later. Of course, this also holds if the initial state already shows one or more discontinuities.

On the other hand, in a suitable definition of a global solution, discontinuities that existed at an instant t_0 can also disappear for $t > t_0$. Examples of this will be presented later. All of these phenomena can occur in scalar problems as well as in systems of conservation laws.

We restrict ourselves within this chapter to only one space variable, but allow the sought solution $V = (v_1, \ldots, v_m)^T$ to consist of more than one component. Hence, we are going to treat the problem

$$\begin{aligned} &\partial_t V + \partial_x f(V) = 0 \quad \text{on} \quad \Omega = \{(x,t) | x \in \mathbb{R},\ t \geq 0\} \ , \\ &V(x,0) = V_0(x) \ , \end{aligned} \tag{2.1}$$

where we will assume for the time being that f is at least one time differentiable.

Example

Another example of this kind – besides (1.14) – is the system of the so-called *shallow-water equations*. These equations describe the shape and the velocity of surface waves for shallow liquids. Here, we consider an idealized situation, where waves of an inviscid incompressible liquid travel along a shallow canal in a direction that coincides with the x-axis, and the flow of liquid particles in the z-direction can be neglected. This idealization appears acceptable if the amplitudes of the surface waves are small compared with the wavelength. The width b of the canal is assumed to be constant, and we are looking for the height $h(x, t)$ of the liquid surface above the bottom of the canal (cf. Fig. 2.1).

Mathematical Models of Fluid Dynamics. R. Ansorge and T. Sonar

ISBN: 978-3-527-40774-3

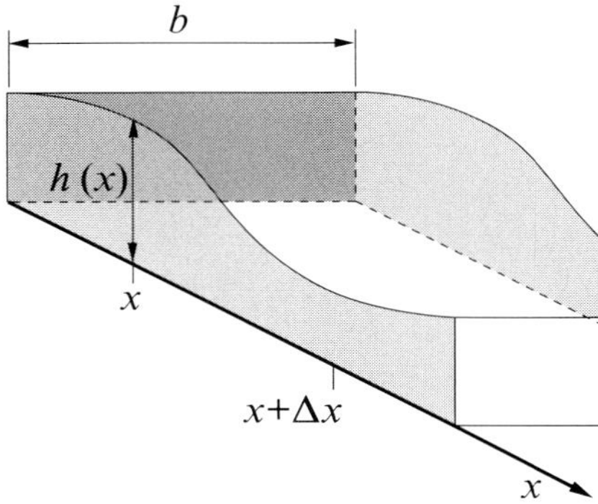

Fig. 2.1 Shallow water flow along a canal.

The mass of the fluid within a segment $x \le \xi \le x+\Delta x$ of the canal at instant t is given by

$$m(t) = b\varrho \int_{x}^{x+\Delta x} h(\xi, t)\, \mathrm{d}\xi \,.$$

With (1.3), if φ is specifically chosen as $\varphi = b\varrho h$, conservation of mass leads to

$$\begin{aligned}\frac{\mathrm{d}}{\mathrm{d}t} m(t) = b\varrho \frac{\mathrm{d}}{\mathrm{d}t} \int_{x}^{x+\Delta x} h(\xi, t)\, \mathrm{d}\xi = b\varrho \int_{x}^{x+\Delta x} \{\partial_t h + \partial_x(uh)\}\, \mathrm{d}\xi \\ = 0 \,, \quad \forall\, (x, \Delta x) \,,\end{aligned}$$

i.e.,

$$\partial_t h + \partial_x(uh) = 0 \,.$$

This replaces the continuity equation in this particular situation.

The momentum of the mass $m(t)$ reads

$$b\varrho \int_{x}^{x+\Delta x} u(\xi, t) h(\xi, t)\, \mathrm{d}\xi \,.$$

The force per length unit arising from the hydrostatic pressure, namely

$$bp = \int_{0}^{h} (h-z)\varrho g\, \mathrm{d}z = \varrho g \frac{h^2}{2} \quad \text{(g: acceleration due to gravity)} \,,$$

acts on the liquid in the segment considered here. The principle of the conservation of momentum then leads via (1.3), now with $\varphi = b\varrho uh$, to

$$\partial_t(\varrho hu) + \partial_x(\varrho hu^2) + \partial_x p = 0\ ,$$

i.e., to

$$\partial_t(hu) + \partial_x\left(hu^2 + \frac{h^2}{2}g\right) = 0\ .$$

This equation turns out to be Euler's equation for our example.

If the continuity equation and the Euler equation are multiplied by g; if $\partial_t(hu)$ is replaced by $\partial_t hu + h\partial_t u$; if a variable $\psi(x,t) := gh(x,t)$ is introduced; if the continuity equation is again taken into account; and finally, if the second equation is divided by ψ, we obtain the *system of shallow-water equations:*

$$\partial_t\begin{pmatrix}\psi\\ u\end{pmatrix} + \partial_x\begin{pmatrix}u\psi\\ \frac{u^2}{2}+\psi\end{pmatrix} = 0\ .$$

Obviously, this is in fact a particular system of the type corresponding to (1.2).

In order to now establish a suitable generalized definition of a *weak solution*, let us temporarily assume that there is a smooth solution to problem (2.1) on Ω. Put this solution into (2.1), then multiply (2.1) by an arbitrary *test function* $\Phi \in C_0^1(\Omega)$. Here, $C_0^1(\Omega)$ is defined to be the set of all functions that are continuously differentiable on Ω and have compact support.

In particular, each of the functions Φ vanishes at the boundary of its support, with the possible exception of the parts of the boundary that belong to the x-axis.

Now we integrate by parts over Ω. In fact, this is only an integration over the compact support of Φ, and hence over a closed, bounded region.

This integration leads to[29)]

$$\int_\Omega \{V\partial_t\Phi + f(V)\partial_x\Phi\}\,\mathrm{d}(x,t) + \int_{-\infty}^{+\infty} V_0(x)\Phi(x,0)\,\mathrm{d}x = 0, \quad \forall\,\Phi \in C_0^1(\Omega)\ . \tag{2.2}$$

Obviously, the left hand side of (2.2) no longer makes sense for only sufficiently smooth functions V; it also holds for all $V \in L_1^{\mathrm{loc}}(\Omega)$, i.e., for all vector-valued functions V whose components can be integrated over compact subsets of Ω in the sense of Lebesgue.

Let us now forget the way we obtained (2.2), and instead ask for functions $V \in L_1^{\mathrm{loc}}(\Omega)$ which fulfill (2.2). These functions will be called *weak solutions* of the original problem (2.1), and these solutions are no longer necessarily smooth.

29) It should be noted that this reformulation of the problem uses the conservation form of the differential equation and cannot be generally applied to arbitrary quasilinear partial differential equations.

On the other hand, our motivation for (2.2) shows that every smooth solution is also a weak solution.

Conversely, every weak solution $\boldsymbol{V}$ which is a smooth function in a neighborhood $D(P_0)$ of a point $P_0 = (x_0, t_0)$ not only fulfills (2.2) but also (2.1) according to the classical understanding. Indeed, let $\boldsymbol{\Phi} \in C_0^1(\Omega)$ be an arbitrary test function with support $\overline{D(P_0)}$. Then

$$\int_{D(P_0)} \{\boldsymbol{V}\partial_t \boldsymbol{\Phi} + \boldsymbol{f}(\boldsymbol{V})\partial_x \boldsymbol{\Phi}\}\, \mathrm{d}(x,t) + \int_{-\infty}^{\infty} \boldsymbol{V}_0(x)\boldsymbol{\Phi}(x,0)\, \mathrm{d}x = 0$$

holds and leads to

$$\int_{D(P_0)} \{\partial_t \boldsymbol{V} \cdot \boldsymbol{\Phi} - \partial_t(\boldsymbol{V}\boldsymbol{\Phi}) + \partial_x \boldsymbol{f}(\boldsymbol{V}) \cdot \boldsymbol{\Phi} - \partial_x(\boldsymbol{f}(\boldsymbol{V}) \cdot \boldsymbol{\Phi})\}\, \mathrm{d}(x,t) - \int_{-\infty}^{\infty} \boldsymbol{V}_0(x)\boldsymbol{\Phi}(x,0)\, \mathrm{d}x = 0\,.$$

By the divergence theorem, the last equation can be written as

$$\int_{D(P_0)} \{\partial_t v_i + \partial_x f_i(\boldsymbol{V})\}\, \boldsymbol{\Phi}\, \mathrm{d}(x,t) - \int_{\partial D(P_0)} \left(\left\langle \begin{pmatrix} f_i(\boldsymbol{V}) \\ v_i \end{pmatrix}, \boldsymbol{n} \right\rangle\right) \boldsymbol{\Phi}\, \mathrm{d}\,(\partial D(P_0)) - \int_{-\infty}^{\infty} v_{0,i}(x)\boldsymbol{\Phi}(x,0)\, \mathrm{d}x = 0$$

$$(i = 1, \ldots, m)\,,$$

where $\boldsymbol{n}$ denotes the outward normal unit vector on the boundary $\partial D(P_0)$ of $D(P_0)$ and where $\langle \cdot, \cdot \rangle$ means the standard $\mathbb{R}^2$-scalar product.

If $\partial D(P_0)$ does not contain points on the x-axis both boundary integrals vanish, because we have $\boldsymbol{\Phi} \equiv 0$ along $\partial D(P_0)$. Otherwise, the first boundary integral turns into an integral over the particular part of the boundary that belongs to the x-axis because $\boldsymbol{\Phi}$ vanishes along the rest of this boundary. In each of these cases we find

$$\int_{D(P_0)} \{\partial_t v_i + \partial_x f_i(\boldsymbol{V})\}\, \boldsymbol{\Phi}\, \mathrm{d}(x,t) - \int_{-\infty}^{\infty} \left\{ \left\langle \begin{pmatrix} f_i(\boldsymbol{V}) \\ v_i \end{pmatrix}, \begin{pmatrix} 0 \\ -1 \end{pmatrix} \right\rangle + v_{0,i} \right\} \boldsymbol{\Phi}(x,0)\, \mathrm{d}x = 0\,,$$

i.e., for $i = 1, \ldots, m$,

$$\int_{D(P_0)} \{\partial_t \boldsymbol{V} + \partial_x \boldsymbol{f}(\boldsymbol{V})\}\, \boldsymbol{\Phi}(x,t)\, \mathrm{d}(x,t) + \int_{-\infty}^{\infty} \{\boldsymbol{V}(x,0) - \boldsymbol{V}_0(x)\}\boldsymbol{\Phi}(x,0)\, \mathrm{d}x = 0\,.$$

Since Φ was arbitrary and V was assumed to be smooth, this can only hold if

$$\partial_t V + \partial_x f(V) = 0$$

on $D(P_0)$, and hence particularly at P_0, and if $V(x_0, 0) = V_0(x_0)$. However, because P_0 was chosen arbitrarily in its neighborhood, our assertion follows.

This concept of a weak solution of (2.1) yields the advantage that the set of admissible solutions can be extended considerably. In particular, discontinuous solutions can be admitted so long as the Lebesgue integrability is not disturbed.

However, there is also a significant disadvantage: the uniqueness of the solution, which was at least guaranteed locally (along with its existence) is lost. There is often more than one weak solution, so a new question arises: how can we select the particular weak solution that answers the original real-world problem from among the set of all of the the weak solutions of (2.1)?

In fact, it is often only possible to describe the solution or weak solution of a differential equation problem approximately, via a numerical procedure. Numerical procedures normally consist of discrete models of the original problem, i.e., in finite-dimensional problems whose solutions are expected to lie in a certain neighborhood of the unknown solution of the original problem. In particular, we expect the approximate solution to converge to this unknown solution if the dimension of the discrete model problem tends to infinity. If convergence of the numerical solution to a weak solution of the original problem can be shown, one must ensure that this weak solution coincides with the relevant solution instead of one of the other weak solutions.[30)]

Let us now study an example where the transition to a weak formulation really does lead to a loss of solution uniqueness.

2.2 Traffic Flow Example with Loss of Uniqueness

As far as sufficiently dense car traffic is concerned, the density ϱ (cars/mile) as well as the flux f (cars/h) in macroscopic traffic flow models are approximately regarded as functions defined on a continuum, e.g., for $a \le x \le b$. The spatial variable x (miles) denotes the positions along a one-way lane which is interpreted as a one-dimensional set, and $0 \le t < \infty$.

If there are no approaches or exits and no crossings along the part of the road under consideration, the conservation of mass (i.e., the constant number of cars) obviously leads within our continuous model to the demand for the validity of the continuity equation

$$\partial_t \varrho + \partial_x f = 0 \tag{2.3}$$

30) Of course, convergence is a basic requirement and must be accompanied by other important properties of the numerical procedure in order to make it an applicable method.

with

$$f = v\varrho\ , \tag{2.4}$$

where $v = v(x, t)$ (miles/h) denotes the speed of the cars at position x and at instant t.

Equation (2.3) must also take an initial condition

$$\varrho(x, 0) = \varrho_0(x) \tag{2.5}$$

with a given function ϱ_0 as well as a spatial boundary condition, since the part of the road under consideration cannot be regarded as a lane of infinite length. Moreover, it should be assumed that – hopefully! – there is no friction between cars.

From experience, ϱ explicitly depends mainly on the speed of the cars:

$$\varrho = \varrho(v)\ . \tag{2.6}$$

This is an empirically given decreasing and strictly monotonic function. Hence, we can also consider v to be a function of ϱ.

Let v_F be the mean value of the individual maximal car velocities if the road in front of a driver is more or less empty. This value is assumed to be a constant, e.g., $v_F = 80$ miles/h.

Obviously, the initial density

$$\varrho_0(x) = \varrho(v(x, 0))$$

is known as soon as the initial speed distribution

$$v(x, 0) = v_0(x) \tag{2.7}$$

is given.

Instead of $\varrho(v)$ or of $v(\varrho)$, the graph of

$$f = v(\varrho) \cdot \varrho =: f(\varrho) \tag{2.8}$$

is often given empirically and turns out to be a strictly concave function called the *fundamental diagram*.

Thus, this traffic flow model leads to an initial value problem for a scalar conservation law of type (2.1), where $\boldsymbol{V}$ consists of just the one component of ϱ such that $\boldsymbol{f}(\boldsymbol{V}) = f(\varrho)$.

Let ϱ^* denote the *jam concentration*, i.e., the density of cars at a standstill in a congestion. This value is also assumed to be a constant, similar to v_F, e.g., $\varrho^* = 400$ cars/ml.

A very simple $f(\varrho)$ model, later used in order to explain certain facts and connections by example, was given by Greenshields:

$$f(\varrho) = v_F \cdot \varrho \cdot \left(1 - \frac{\varrho}{\varrho^*}\right)\ . \tag{2.9}$$

Thus, the flux vanishes – in other words, there is (almost) no traffic – if the traffic density vanishes or if the road users are at a standstill within a traffic jam.

Equation (2.9) is a concave parabola that has its maximum at $(\frac{\varrho^*}{2}, \frac{v_F \varrho^*}{4})$, and it leads – together with (2.4) – to the relation

$$\varrho = \varrho^* \left(1 - \frac{v}{v_F}\right) .$$

Hence, in the case of Greenshields' model, the graph of $\varrho(v)$ is found to be a straight line.

The more general model for arbitrary fluxes is due to Lighthill & Whitham.[31)]

Using Greenshields' model, we are going to begin at the start of the traffic jam, i.e.,

$$\varrho_0(x) = \begin{cases} \varrho^* & \text{for} \quad x < 0 \\ 0 & \text{for} \quad x \geq 0 \end{cases} \tag{2.10}$$

for $t_0 = 0$. The start of the traffic jam (red traffic light) is located at $x = 0$, and we identify the forward direction of the traffic with the positive direction along the x-axis.

Remark

Conservation law problems with initial values that are constant on the left hand side of a discontinuity as well as on its right hand side are called *Riemann problems.*[32)]

With (2.9) and (1.17), the characteristics are given by

$$\frac{1}{v_F}(x - x_0) = \begin{cases} -t & \text{for} \quad x_0 < 0 \\ t & \text{for} \quad x_0 \geq 0 \end{cases} .$$

Obviously, the characteristics do not intersect for $t > 0$, so we immediately find that

$$\varrho(x, t) = \begin{cases} \varrho^* & \text{for} \quad x < -v_F t \\ 0 & \text{for} \quad x \geq v_F t \, . \end{cases}$$

Thus, the question of how the solution will look within the hatched region of

31) M.J. Lighthill, C.B. Whitham: Proc. Roy. Soc. **A 229** (1955) 317–345

31) Bernhard Riemann (1826–1866); Göttingen

Fig. 2.2, i.e., for

$$-v_F t < x < v_F t, \quad t > 0 ,$$

is open. It can easily be verified that

$$\varrho_1(x,t) := \begin{cases} \varrho^* & \text{for} \quad x < 0 \\ 0 & \text{for} \quad x \geq 0 \end{cases} \qquad \forall\, t \geq 0 \tag{2.11}$$

is a weak solution of

$$\partial_t \varrho + f_v \cdot \partial_x \left(\varrho \left(1 - \frac{\varrho}{\varrho^*} \right) \right) = 0 ,$$

i.e., it is of the Greenshields type of (2.3) together with the initial condition (2.10), because, if realized for our particular flux f, the left hand side of (2.2) leads to

$$\int_0^\infty \int_{-\infty}^\infty \left[\partial_t \Phi \varrho_1 + \partial_x \Phi v_F \varrho_1 \left(1 - \frac{\varrho_1}{\varrho^*} \right) \right] \mathrm{d}(x,t) + \int_{-\infty}^\infty \Phi(x,0) \varrho_0(x) \, \mathrm{d}x$$

$$= \int_0^\infty \int_{-\infty}^0 \varrho^* \partial_t \Phi \, \mathrm{d}(x,t) + \int_{-\infty}^0 \Phi(x,0) \varrho^* \, \mathrm{d}x$$

$$= \varrho^* \int_{-\infty}^0 \left\{ \int_0^\infty \partial_t \Phi \, \mathrm{d}t + \Phi(x,0) \right\} \mathrm{d}x = \varrho^* \int_{-\infty}^0 \{ \Phi(x,0) - \Phi(x,0) \} \, \mathrm{d}x = 0 .$$

In other words, (2.1) is a solution which keeps its initial state. The drivers do not move even though the traffic light has switched from red to green. This solution describes a special situation where the state along the left hand boundary of the hatched region of Fig. 2.2, i.e., along the characteristic $t = -\frac{x}{v_F}$, is connected with the right hand boundary $t = \frac{x}{v_F}$ across a line of discontinuity.

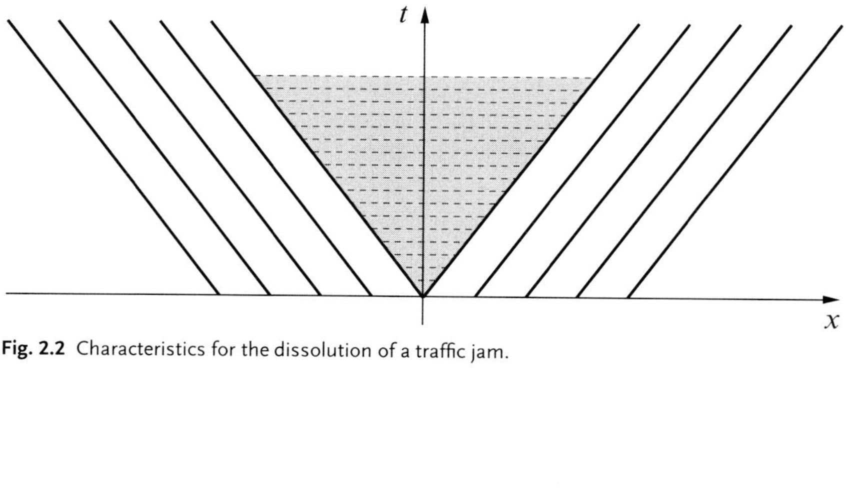

Fig. 2.2 Characteristics for the dissolution of a traffic jam.

Of course, drivers do not behave this way, so (2.11) is certainly an unreasonable and hence irrelevant solution. The drivers' *ride impulse* has not yet been introduced into the model.

However, there is another weak solution to our problem, namely

$$\varrho_2(x,t) := \begin{cases} \varrho^* & \text{for} \quad x < -v_F t\,, \quad t \geq 0 \\ \varrho^* \dfrac{v_F t - x}{2 v_F t} & \text{for} \quad -v_F t \leq x \leq v_F t\,, \quad t > 0\,, \\ 0 & \text{for} \quad x > v_F t\,, \quad t \geq 0 \end{cases} \tag{2.12}$$

which can also easily be verified by elementary integration procedures. Within this solution, the states along the left and the right hand sides of the hatched region of Fig. 2.2 are continuously connected. In this case, the density decreases, so this type of a solution is called a *rarefaction wave*.

The head of the traffic jam, i.e., the the front end of the chain of cars at a standstill, withdraws against the direction of movement of the traffic with speed $-v_F$, whereas the cars in front move with speed $+v_F$. Obviously, (2.12) coincides with our experiences within the bounds of Greenshields' idealizations. It must therefore be regarded as the particular weak solution that fits the circumstances of the given real situation.[33)]

Of course, loss of uniqueness can also occur if systems of conservation laws are involved instead of scalar situations.

As another application of Greenshields' traffic flow model, let us consider the propagation of the end of the traffic jam opposite to the direction of the movement.

We start from an initial situation at the instant $t = 0$, where the traffic shows maximal flux on the left of $x = 0$, i.e., $\varrho = \frac{\varrho^*}{2}$, whereas the cars on the right of $x = 1$ have already stopped, i.e., $\varrho = \varrho^*$, and where we have a continuous linear transition in-between:

$$\varrho_0(x) = \begin{cases} \dfrac{\varrho^*}{2} & \text{for} \quad x < 0 \\ \dfrac{\varrho^*}{2}(1+x) & \text{for} \quad 0 \leq x \leq 1\,. \\ \varrho^* & \text{for} \quad x > 1 \end{cases} \tag{2.13}$$

The characteristics then turn out to be

$$\left.\begin{aligned} x &= x_0 && \text{for} \quad x_0 \leq 0 \\ t &= \frac{1}{v_F}\left(1 - \frac{x}{x_0}\right) && \text{for} \quad 0 < x_0 \leq 1 \\ t &= -\frac{x - x_0}{v_F} && \text{for} \quad x_0 > 1 \end{aligned}\right\}. \tag{2.14}$$

33) This is also an example of a situation mentioned earlier, namely that a solution that is continuous for $t > 0$ can develop from a discontinuous initial function.

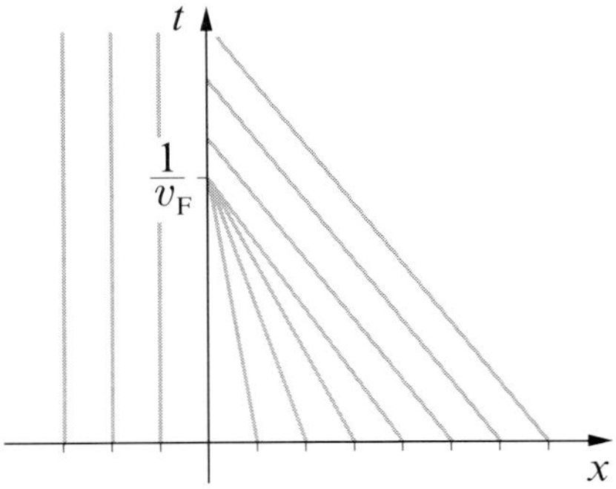

Fig. 2.3 Characteristics at the creation of a traffic jam.

This solution $\varrho(x, t)$ maintains its continuity for the time $0 \le t < \frac{1}{v_F}$. Then, a discontinuity suddenly occurs at $t = \frac{1}{v_F}$, and this discontinuity moves for $t > \frac{1}{v_F}$ along an as-yet unknown curve Γ. However, we know that this curve occurs in the upper left of the plane because the characteristics arising from the x-axis on the right of $x_0 = 1$ undoubtedly intersect with the characteristics arising from $x_0 < 0$. As soon as Γ becomes known, the values of ϱ given along the characteristics make it possible to explicitly compute a solution $\varrho(x, t)$. It is discontinuous along Γ but still an element of L_1^{loc}, and thus actually a weak solution.

The drivers do not expect such a discontinuity to crop up, and this is often one of the reasons for a rear-end collision.

One of the examples in the next section will show how Γ can be determined.

2.3 The Rankine–Hugoniot Condition

We are going to treat the particular case where a certain discontinuity of a weak solution $\boldsymbol{V}$ of (2.1) moves continuously along the x-axis. In other words, the set of points of Ω where these discontinuities occur is a continuous curve Γ in Ω. In this case, it may be that all of the components of $\boldsymbol{V}$ along this path are discontinuous, or that this is only true of some of them. In gas dynamics, often the density is particularly disturbed, resulting in shocks.

Let $P_0 = (x_0, t_0) \in \Gamma$ with $t_0 > 0$, and assume that a bounded neighborhood $\overline{D(P_0)} \subset \Omega$ of P exists which does not contain points on the x-axis and where $\boldsymbol{V}$ is smooth outside Γ.[34)]

We decompose $\overline{D(P_0)}$ into two parts, $\overline{D_1}$ and $\overline{D_2}$ (cf. Fig. 2.4), each of which we are going to treat as closed domains.

Taking the smoothness of $\boldsymbol{V}$ in the open domains D_1 and D_2 into account, for a particular $\Phi \in C_0^1(\Omega)$ (the compact support of which coincides with $\overline{D(P_0)} =$

34) The considerations to be formulated here can easily be generalized in order to fit more complicated situations too.

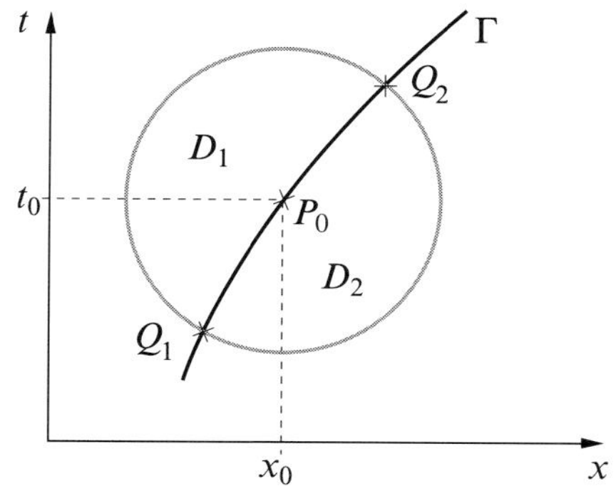

Fig. 2.4 Sketch to explain the proof of the Rankine–Hugoniot condition.

$\overline{D_1 \cup D_2}$), we find on $\overline{D_1}$ and on $\overline{D_2}$ the relation

$$
\begin{aligned}
&-\int\limits_{\Omega} \left[V\partial_t \Phi + f(V)\partial_x \Phi \right] \mathrm{d}(x,t) - \int\limits_{-\infty}^{\infty} V_0(x)\Phi(x,0)\,\mathrm{d}x \\
&= -\int\limits_{D_1} \left[\partial_t(V\Phi) + \partial_x(f(V)\Phi) \right] \mathrm{d}(x,t) + \int\limits_{D_1} \left[\partial_t V\Phi + \partial_x f(V)\Phi \right] \mathrm{d}(x,t) \\
&\quad -\int\limits_{D_2} \left[\partial_t(V\Phi) + \partial_x(f(V)\Phi) \right] \mathrm{d}(x,t) + \int\limits_{D_2} \left[\partial_t V_t\Phi + \partial_x f(V)\Phi \right] \mathrm{d}(x,t) \\
&= 0\,.
\end{aligned}
$$

Together with the validity of (2.1) in D_1 and in D_2,

$$
\int\limits_{D_1} \left[\partial_t(V\Phi) + \partial_x(f(V)\Phi) \right] \mathrm{d}(x,t) + \int\limits_{D_2} \left[\partial_t(V\Phi) + \partial_x(f(V)\Phi) \right] \mathrm{d}(x,t) = 0\,.
$$

If applied to each of the components of both integrals, the divergence theorem then leads to

$$
\int\limits_{\partial D_1} \{(V\Phi)\,\mathrm{d}x - (f(V)\Phi)\,\mathrm{d}t\} + \int\limits_{\partial D_2} \{(V\Phi)\,\mathrm{d}x - (f(V)\Phi)\,\mathrm{d}t\} = 0\,.
$$

Because Φ vanishes at the boundary of its support, i.e., at the boundary of D, we find

$$
\int\limits_{Q_1}^{Q_2} \{(V_{\mathrm{L}}\Phi)\,\mathrm{d}x - (f(V_{\mathrm{L}})\Phi)\,\mathrm{d}t\} - \int\limits_{Q_1}^{Q_2} \{(V_{\mathrm{R}}\Phi)\,\mathrm{d}x - (f(V_{\mathrm{R}})\Phi)\,\mathrm{d}t\} = 0\,,
$$

where V_{L} denotes the left hand limits of V along Γ if the values of x approach Γ from the left for every fixed value of t, i.e., $V_{\mathrm{L}}(x,t) = V(x-0,t)$ for $(x,t) \in \Gamma$. Analogously, V_{R} denotes the values of V if we approach Γ from the right. The integrals should be considered line integrals along Γ.

Let

$$[V] := V_L - V_R \ , \quad [f] := f_L - f_R := f(V_L) - f(V_R)$$

be the *jumps*.

Then, obviously,

$$\int_{Q_1}^{Q_2} \left\{ [V] \frac{dx}{dt} - [f] \right\} \Phi \, dt = 0 \ ,$$

where $\frac{dx}{dt} =: \hat{v}$ denotes the velocity with which the discontinuity moves along the x-axis. Since the test function Φ with support $\overline{D(P_0)}$ is arbitrary, and because P_0 is an arbitrarily chosen point on Γ, the so-called *Rankine–Hugoniot condition*

$$[V]\hat{v} = [f] \tag{2.15}$$

is found to hold along Γ component-wise under the conditions that were assumed to hold.

Equation (2.15) is also called the *jump condition*, and for every weak solution V there is a particular velocity $\hat{v} = \hat{v}(V)$.

Example

If (2.15) is applied to our one-dimensional gas flow where

$$V = \begin{pmatrix} \varrho \\ q \\ E \end{pmatrix} \ , \quad f = \begin{pmatrix} q \\ \frac{1}{\varrho} q^2 + p \\ \frac{E + p}{\varrho} q \end{pmatrix} \ , \tag{2.16}$$

and if $q = \varrho u$, with the flow velocity u taken into account, we find from the first component of the jump condition

$$(\varrho_L - \varrho_R)\hat{v} = q_L - q_R = \varrho_L u_L - \varrho_R u_R$$

i.e.,

$$\varrho_L(\hat{v} - u_L) = \varrho_R(\hat{v} - u_R) \ . \tag{2.17}$$

The second component leads to

$$(q_L - q_R)\hat{v} = \frac{q_L^2}{\varrho_L} - \frac{q_R^2}{\varrho_R} + p_L - p_R \ ,$$

i.e.,

$$\varrho_L u_L(\hat{v} - u_L) = \varrho_R u_R(\hat{v} - u_R) + p_L - p_R \,, \tag{2.18}$$

and the third component gives

$$(E_L - E_R)\hat{v} = E_L \frac{q_L}{\varrho_L} - E_R \frac{q_R}{\varrho_R} + p_L \frac{q_L}{\varrho_L} - p_R \frac{q_R}{\varrho_R} \,,$$

i.e.,

$$E_L(\hat{v} - u_L) = E_R(\hat{v} - u_R) + p_L u_L - p_R u_R \,. \tag{2.19}$$

If

$$u_L = u_R = \hat{v} \,, \tag{2.20}$$

(2.17) is fulfilled, and (2.18) then gives $p_L = p_R$, such that (2.19) is also realized.
Hence, in this case, the discontinuity "swims" with the flow; this means that there are no gas particles that cross the point of discontinuity, though $\varrho_L = \varrho_R$ or $E_L = E_R$ do not necessarily hold.

The situation described by

$$q_L \varrho_R = q_R \varrho_L \,, \quad \varrho_L \neq \varrho_R \tag{2.21}$$

is called *contact discontinuity*.

Otherwise, $u_L \neq \hat{v}$, $u_R \neq \hat{v}$ show that there are gas particles which cross the discontinuity, because in this case (2.17) shows that $\hat{v} - u_L$ and $\hat{v} - u_R$ are both positive or both negative.[35] This type of discontinuity is called a *shock*, and the curve Γ is also called a shock or a shock curve. The particular side of the shock where particles that have already crossed the shock are situated is called the *back side of the shock*, while the other part is called the *front side*. If, for instance, $\hat{v} > u_R > 0$, the right hand side is the front side and the left hand side is the back side.

Assume a weak solution to be smooth outside the discontinuity curve Γ, hence to be outside Γ also a solution of (2.1). The weak formulation (2.2) of the problem under consideration is then equivalent to the demand that the Rankine–Hugoniot condition is fulfilled along Γ and that the original version of the initial value problem (2.1) is simultaneously fulfilled outside Γ too. This equivalence follows from the already proven fact that weak solutions are also classical solutions in domains where these weak solutions are smooth.

In the case of a linear system

$$\partial_t \boldsymbol{V} + \partial_x(\boldsymbol{A}\boldsymbol{V}) = 0 \,,$$

the Rankine–Hugoniot condition leads to

$$[\boldsymbol{V}] \cdot \hat{v} = [\boldsymbol{A}\boldsymbol{V}_L - \boldsymbol{A}\boldsymbol{V}_R] = \boldsymbol{A}[\boldsymbol{V}] \,.$$

35) We omit the unnatural case $\varrho < 0$.

Thus, $\hat{v}$ is an eigenvalue of $\boldsymbol{A}$ as far as a discontinuity of the initial function occurs, i.e., as far as $[\boldsymbol{V}] \neq 0$ is found to hold for $t = 0$ (and thus also for increasing values of t). This leads toq

$$\dot{x} = \lambda_k \quad \text{for a particular} \quad k$$

since $\dot{x}(t) = \hat{v}$ is the differential equation for Γ.

However, because of (1.18), this is also the differential equation for the k-th set of characteristic curves, so that discontinuities move along characteristics as far as linear problems are concerned.

Let us wind up this section by determining the Γ in Greenshields' traffic flow model for the case of a traffic jam that increases in the direction opposite to the flow direction.

Because of the relation

$$f_{\mathrm{L}} - f_{\mathrm{R}} = v_{\mathrm{F}} \left(\varrho_{\mathrm{L}} - \frac{\varrho_{\mathrm{L}}^2}{\varrho^*} - \varrho_{\mathrm{R}} + \frac{\varrho_{\mathrm{R}}^2}{\varrho^*} \right)$$

presented in the Greenshields model, and taking (2.15) into account, Γ has to be determined by the differential equation

$$\dot{x} = \frac{f_{\mathrm{L}} - f_{\mathrm{R}}}{\varrho_{\mathrm{L}} - \varrho_{\mathrm{R}}} = v_{\mathrm{F}} \left(1 - \frac{\varrho_{\mathrm{L}} + \varrho_{\mathrm{R}}}{\varrho^*} \right) .$$

From Fig. 2.3, the initial condition is

$$x \left(\frac{1}{v_{\mathrm{F}}} \right) = 0 . \tag{2.22}$$

Moreover, Fig. 2.3 shows that

$$\varrho = \varrho^* \qquad \text{for} \qquad t > \frac{1}{v_{\mathrm{F}}}$$

along the characteristics from the right hand side, and this does not depend on the location of Γ. Along the vertical characteristics from the left side, we analogously find that

$$\varrho = \frac{\varrho^*}{2} \qquad \text{for} \qquad t > \frac{1}{v_{\mathrm{F}}} .$$

Hence $\varrho_{\mathrm{L}} = \frac{\varrho^*}{2}$, and therefore

$$\dot{x} = -\frac{v_{\mathrm{F}}}{2} .$$

Together with (2.22), the shock is found to be the straight line

$$t = \frac{1}{v_{\mathrm{F}}} (1 - 2x) , \quad x \leq 0 .$$

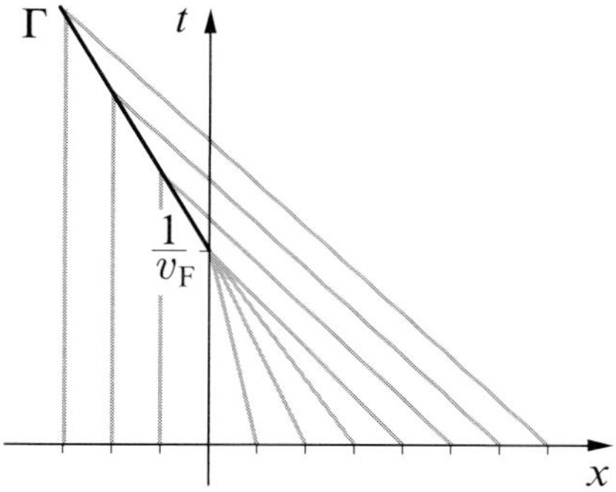

Fig. 2.5 Shock when a traffic jam increases.

Knowing Γ and the values of ϱ along the characteristic curves, we can now immediately construct a weak solution to our traffic flow example. In particular, the propagation of the end of the congestion against the direction of travel of the still-moving cars can be evaluated:

$$\varrho(x,t) = \begin{cases} \dfrac{\varrho^*}{2} & \text{for} \quad 0 \le t < \dfrac{1}{v_F}(1-2x)\,, \quad -\infty < x \le 0 \\ \dfrac{\varrho^*}{2}\left(1 + \dfrac{x}{1 - v_F t}\right) & \text{for} \quad 0 \le t < \dfrac{1-x}{v_F}\,, \quad 0 \le x < 1 \\ \varrho^* & \text{otherwise}\,. \end{cases} \tag{2.23}$$

This solution shows that, under the assumptions of the Greenshields idealizations, and for the particular initial situation under consideration, that the end of the chain of stationary cars originally located at $x = 1$ will be located at $x = 0$ after $\frac{1}{v_F}$ time units. From this moment on, the end of the traffic jam moves with the *congestion velocity* $\frac{1}{2}v_F$ against the direction of the still-moving cars, and the transition from the density $\frac{1}{2}v_F$ of the moving traffic to the density ϱ^* of the set of cars that have already come to a halt occurs very suddenly, at least from the driver's point of view.

With respect to the first part of our Lighthill–Whitham–Greenshields traffic flow model,[36] as well as generally speaking, a question now arises: how do we characterize the particular solution that can be used to answer the real-world problem underlying the mathematical model for the non-unique weak solutions of (2.1)?

We are going to address this question in the next chapter.

36) A more realistic but still continuous model can be found in Section 4.3.

3
Entropy Conditions

3.1
Entropy in the Case of an Ideal Fluid

The loss of uniqueness caused by adopting the concept of weak solutions leads to the need to formulate a criterion by which the physically relevant weak solution can be selected from the set of all of the weak solutions. We therefore require an additional constraint that will characterize this particular weak solution.

In order to work out how to do this, let us look again at the important application of conservation laws in order to describe the dynamics of ideal gases. Besides the state equation, we took into account the conservation of mass, the conservation of momentum and the conservation of energy. The conservation of energy was formulated by the first law of thermodynamics, but what we have not considered yet is the second law, and its assertion about the behavior of the state variable

$$S = \text{entropy}/\text{mass} .$$

Without going into detail, we will mention that the entropy describes a measure of the probability that a particular physical state exists (*Boltzmann statistics* in thermodynamics[37]).

The *second law of thermodynamics*[38] can be stated as follows:

If there is a closed physical system without any supply of energy from outside, every physical process inside the system takes a course such that the entropy does not decrease,[39] *and it increases if possible.*[40]

Figure 3.1 shows an example of the validity of the second law, namely a shock tube divided by a membrane into two parts, one of which is filled with a gas while the other is evacuated. When the membrane is taken away, some of the gas will begin to move into the evacuated part of the tube, even though all of the conservation principles would also hold if the gas particles maintained their positions (in fact, the Euler equations would also hold in this case). However this is obviously not

37) Ludwig Boltzmann (1844–1906); Graz, Vienna, Leipzig, Vienna

38) R.J.E. Clausius (1822–1888); Zurich, Würzburg, Bonn

39) See, e.g., R. Ansorge, T. Sonar: ZAMM 77 (1997) 803–821

40) Irreversible processes

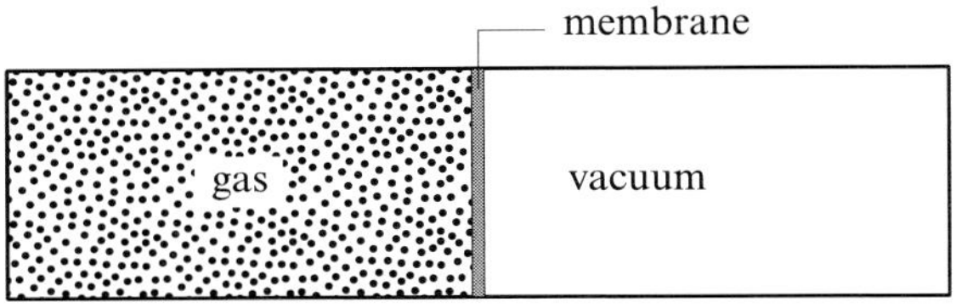

Fig. 3.1 Example of increasing entropy.

a physically relevant solution, as everybody knows from experience.[41] The reason for this is that the equidistribution of the particles throughout the whole tube is a more probable state than the initial situation.

In other words, we expect the second law to be a model for the formulation of a constraint that allows us to pick out the relevant solution from the set of weak solutions of a conservation law system, even when the problems are from other fields of application, such as economics. In any case, such a constraint will be called an *entropy condition* and the selected solution will be called an *entropy solution*; hopefully the only one obtained from a proof of uniqueness theorem.

Generalizations of the second law in order to find a constraint that works for more than just physical tasks have been derived by several authors, particularly Oleinik and Lax, who initially tackled this question in relation to scalar conservation laws.

To understand their ideas, let us look at the situation in gas dynamics; for convenience, we consider a one-dimensional flow.

It was stated in Section 1.1 that each of the state variables can be uniquely expressed in terms of two of the other state variables via a so-called equation of state; (1.9) was a particular example.

One result of the theory of thermodynamics leads to the relation

$$\mathrm{d}S = \frac{\mathrm{d}\varepsilon + p\,\mathrm{d}\tilde{W}}{T} \tag{3.1}$$

where $\tilde{W}$ denotes the specific volume $\tilde{W} := \frac{1}{\varrho}$ (volume of mass 1).

If we look for the connection between the three state variables ε, S and $\varrho = \frac{1}{\tilde{W}}$, the equation of state belonging to these variables leads to

$$\mathrm{d}\varepsilon = \partial_S \varepsilon\,\mathrm{d}S + \partial_{\tilde{W}} \varepsilon\,\mathrm{d}\tilde{W}\,,$$

and hence with (3.1) to

$$T\,\mathrm{d}S = \partial_S \varepsilon\,\mathrm{d}S + \partial_{\tilde{W}} \varepsilon\,\mathrm{d}\tilde{W} + p\,\mathrm{d}\tilde{W} = \partial_S \varepsilon\,\mathrm{d}S + (\partial_{\tilde{W}} \varepsilon + p)\,\mathrm{d}\tilde{W}\,,$$

such that

$$\partial_S \varepsilon = T\,, \qquad \partial_{\tilde{W}} \varepsilon = -p \tag{3.2}$$

follows from the linear independence of $\mathrm{d}S$ and $\mathrm{d}\tilde{W}$.

41) By the way, this is analogous to the situation where a traffic light switches from red to green but the drivers do not move.

Because of

$$\partial_t \varepsilon = \partial_S \varepsilon \partial_t S + \partial_{\tilde{W}} \varepsilon \partial_t \tilde{W} = T \partial_t S - p \partial_t \tilde{W} ,$$

the relation

$$\left.\begin{aligned} T\partial_t S &= \partial_t \varepsilon + p \partial_t \tilde{W} = \partial_t \varepsilon - p \frac{\partial_t \varrho}{\varrho^2} \\ T\partial_x S &= \qquad\qquad\quad\; \partial_x \varepsilon - p \frac{\partial_x \varrho}{\varrho^2} \, . \end{aligned}\right\} \tag{3.3}$$

follows. Analogously:

Denoting the entropy per volume as $s := \varrho S$, the quantity $\partial_t s + \partial_x(us)$ can be expressed as

$$\begin{aligned} \partial_t s + \partial_x(us) &= \partial_t(\varrho S) + \partial_x((\varrho u) \cdot S) = \partial_t \varrho S + \varrho \partial_t S + \partial_x(\varrho u) S + \varrho u \partial_x S \\ &= [\partial_t \varrho + \partial_x(\varrho u)] \cdot S + \varrho[\partial_t S + u \partial_x S] \, . \end{aligned}$$

The first term on the right hand side vanishes because of the continuity equation, and taking (3.3) into account, the last equation leads to

$$\partial_t s + \partial_x(us) = \frac{\varrho}{T} \left\{ \left(\partial_t \varepsilon - p \frac{\partial_t \varrho}{\varrho^2} \right) + u \left(\partial_x \varepsilon - p \frac{\partial_x \varrho}{\varrho^2} \right) \right\} . \tag{3.4}$$

The conservation law (1.7) concerning the specific total energy E, and formulated for the particular situation of a one-dimensional flow, reads

$$\partial_t E + \partial_x \left(\frac{E + p}{\varrho} q \right) = 0$$

or, with $E = \varrho \varepsilon + \frac{\varrho}{2} u^2$ (reformulated using primitive variables),

$$\begin{aligned} &\partial_t \varrho \varepsilon + \varrho \partial_t \varepsilon + \frac{1}{2} \partial_t \varrho u^2 + \varrho u \partial_t u + u \left\{ \partial_x p + \partial_x \varrho \varepsilon + \varrho \partial_x \varepsilon + \frac{1}{2} \partial_x \varrho u^2 + \varrho u \partial_x u \right\} \\ &+ \left(\varrho \varepsilon + \frac{\varrho}{2} u^2 + p \right) \partial_x u = 0 \, , \end{aligned}$$

i.e.,

$$\begin{aligned} (\partial_t \varrho + \partial_x(\varrho u)) \left(\varepsilon + \frac{u^2}{2} \right) + \varrho \left(\partial_t \varepsilon + u \partial_x \varepsilon + \frac{p}{\varrho} \partial_x u \right) \\ + \varrho u \left(\partial_t u + \frac{\partial_x p}{\varrho} + u \partial_x u \right) = 0 \, . \end{aligned} \tag{3.5}$$

Also, in this equation, the first term vanishes because of the continuity equation. The second conservation law, namely $\partial_t q + \partial_x \left(\frac{q^2}{\varrho} + p \right) = 0$, also rewritten using primitive variables, leads to

$$\varrho \left(\partial_t u + \frac{\partial_x p}{\varrho} + u \partial_x u \right) + u(\partial_t \varrho + \partial_x(\varrho u)) = 0 \, ,$$

so that the third term on the left hand side of (3.5) also disappears, again through the use of the continuity equation. Thus, after dividing by ϱ,

$$\partial_t \varepsilon + u \partial_x \varepsilon + \frac{p}{\varrho} \partial_x u = 0 \,, \tag{3.6}$$

remains, so the term inside the outer braces in (3.4) becomes

$$-\frac{p}{\varrho} \partial_x u - \frac{p}{\varrho} \frac{\partial_t \varrho}{\varrho} - \frac{p}{\varrho} \frac{u \partial_x \varrho}{\varrho} = -\frac{p}{\varrho^2} (\partial_t \varrho + \partial_x(\varrho u)) = 0 \,, \tag{3.7}$$

again because of the continuity equation.

Hence, s also fulfills a conservation law:

$$\partial_t s + \partial_x(us) = 0 \,, \tag{3.8}$$

and this holds *automatically* if the Euler equations are fulfilled.

However, we have to take into account that this result requires that the differentiability properties we used to derive (3.8) are ensured. Thus, if the flow is smooth, Section 1.1 shows that the total entropy

$$\int_{W(t)} s \, \mathrm{d}(x, y, z)$$

of an volume $W(t)$ of the fluid selected arbitrarily at an arbitrary instant t stays constant with respect to time. This is known as *isentropic flow*.

Equation (3.8) cannot be expected to hold if the smoothness assumptions are not fulfilled. What can be expected, however, because of the second main theorem, is the validity of the inequality

$$\partial_t s + \partial_x(us) \geq 0 \tag{3.9}$$

in its weak form, i.e.,

$$\int_{\Omega} (s \partial_t \Phi + us \partial_x \Phi) \, \mathrm{d}(x, t) + \int_{\mathbb{R}} s(x, 0) \Phi(x, 0) \, \mathrm{d}x \leq 0$$
$$\forall \, \Phi \in C_0^1(\Omega) \quad \text{with} \quad \Phi \geq 0 \,. \tag{3.10}$$

Formally, and because it is an inequality, (3.9) was multiplied by *nonnegative* test functions, then integrated by parts, and finally this "derivation" of (3.10) should be forgotten.

Equation (3.10) is the entropy condition for the flow under consideration, and the "derivation" shows that it is indeed satisfied automatically provided that the original conservation laws are fulfilled according to the classical understanding. This follows from the fact that (3.8) was a direct consequence of the Euler equations.

If the weak solution $\boldsymbol{V}$ of the Euler equations is discontinuous, and thus not necessarily unique, (3.10) is actually an *additional* condition.

Remark

We mention without proof that the physical entropy per volume $-s = -s(\boldsymbol{V})$ in gas dynamics is strictly convex.

3.2
Generalization of the Entropy Condition

Lax[42] provided a definition of an entropy solution for a general system of conservation laws (2.1) that can, in the case of one spatial variable and by imitating the gas dynamics situation, be formulated as follows:

Definition

A weak solution $V = (v_1, \ldots, v_m)^T$ of the system (2.1), i.e., a solution of (2.2), is called an *entropy solution* of this system if there is a scalar and strictly convex function $\tilde{S} = \tilde{S}(V)$ as well as a scalar function $\tilde{F} = \tilde{F}(V)$ belonging to $\tilde{S}$ such that, in every domain where V is smooth, (2.1) leads *automatically* to the validity of

$$\partial_t \tilde{S}(V) + \partial_x \tilde{F}(V) = 0 \,, \tag{3.11}$$

and such that the *entropy condition*

$$\int\limits_{\Omega} \left\{ \tilde{S}(V(x,t))\partial_t \Phi(x,t) + \tilde{F}(V(x,t))\partial_x \Phi(x,t) \right\} \, \mathrm{d}(x,t) + \int\limits_{\mathbb{R}} \tilde{S}(V_0(x))\Phi(x,0)\, \mathrm{d}x \geq 0 \,, \quad \forall \Phi \in C_0^1(\Omega) \,, \ \Phi \geq 0 \tag{3.12}$$

is respected in the non-smooth situation.

Remarks

- $\tilde{S}$ is often also called the *entropy functional* and $\tilde{F}$ is called the *entropy flux*
- In gas dynamics we have $\tilde{S} = -s$
- On occasion we will talk more specifically of Lax entropy solutions, because some other definitions will also be given later
- (3.12) is often abbreviated by the formulation:

  $$\partial_t \tilde{S}(V) + \partial_x \tilde{F}(V) \leq 0 \ \ \textit{holds weakly}$$

After choosing $\tilde{S}$, $\tilde{F}$ is already more or less determined by (3.11). Especially in the scalar case $m = 1 (V = v, f = f)$, (3.11) leads for smooth solutions v to

$$\tilde{S}'(v) \cdot \partial_t v + \tilde{F}'(v)\partial_x v = 0 \,,$$

and therefore with (2.1) to

$$\tilde{S}'(v) \left\{ -\partial_x f(v) \right\} + \tilde{F}'(v)\partial_x v = 0 \,,$$

42) P. Lax in: E. Zarantonello (ed.): *Contributions to nonlinear functional analysis.* New York: Academic 1971

i.e., to

$$\left[-\tilde{S}'(v)f'(v) + \tilde{F}'(v)\right] \partial_x v = 0 \, .$$

Because this must hold automatically, independent of the choice of $v_o(x)$,[43] the relation

$$\tilde{F}'(v) = \tilde{S}'(v)f'(v)$$

must be fulfilled, i.e.,

$$\tilde{F}(v) = \int_0^v \tilde{S}'(\alpha)f'(\alpha)\, d\alpha + \text{const} \, . \tag{3.13}$$

Only the constant remains arbitrary.

If $m > 1$ and the function V is smooth, (3.11) yields

$$\langle \nabla_V \tilde{S}, \partial_t V \rangle + \langle \nabla_V \tilde{F}, \partial_x V \rangle = 0 \, ,$$

where

$$\nabla_V = \left(\partial_{v_1}, \partial_{v_2}, \ldots, \partial_{v_m}\right)^T \, .$$

By analogy with $m = 1$, putting (2.1) into this equation, we find

$$-\langle \nabla_V \tilde{S}, \boldsymbol{J}\boldsymbol{f}(V)\partial_x V \rangle + \langle \nabla_V \tilde{F}, \partial_x V \rangle = 0 \, , \text{ i.e.}$$
$$(\partial_x V)^T \left(\nabla_V \tilde{F} - (\boldsymbol{J}\boldsymbol{f}(V))^T \nabla_V \tilde{S}\right) = 0 \, .$$

Again, because this relation must hold for any choice of V_0, $\tilde{S}$ and $\tilde{F}$ have to be connected by

$$\nabla_V \tilde{F} - (\boldsymbol{J}\boldsymbol{f}(V))^T \nabla_V \tilde{S} = 0 \, , \tag{3.14}$$

and this is a system of m differential equations for only two functions $\tilde{S}$; $\tilde{F}$ (to be determined) that depends on the m independent variables $v_1, \ldots, v_m$.

Hence, (3.14) does not necessarily have a solution when $m > 2$,[44] so the definition of an entropy condition used so far must be modified appropriately. We shall come back to this question in Section 3.3.

However, if $\boldsymbol{f}(V)$ is a gradient too, i.e.,

$$\boldsymbol{f}(V) = \nabla_V k(V)$$

43) In particular, for initial functions that generate smooth solutions

44) As far as gas dynamics is concerned, a suitable entropy flux belonging to $\tilde{S} = -s$ exists, though $3 \le m \le 5$.

with a scalar function $k(V)$, the particular choice of $\tilde{S}(V) = \frac{1}{2}\langle V, V\rangle$ turns (3.14) into the following system of differential equations:

$$\partial_{v_i}\tilde{F} = \sum_{\nu=1}^{m} \partial_{v_\nu v_i} k v_\nu = \partial_{v_i}\left(\sum_{\nu=1}^{m} \partial_{v_\nu} k v_\nu\right) - \partial_{v_i} k \qquad (i = 1, \ldots, m)\ ,$$

and this system can be satisfied by several solutions, e.g., by the particular entropy flux

$$\tilde{F}(V) = \langle f(V), V\rangle - k(V) + \text{const}\ . \tag{3.15}$$

A straightforward formulation of these considerations for $m = 1$ leads directly to the particular pair

$$\tilde{S}(v) = \frac{1}{2}v^2\ , \quad \tilde{F}(v) = \int_0^v f'(\alpha)\alpha\, \mathrm{d}\alpha + \text{const}\ , \tag{3.16}$$

but an infinite number of other suitable pairs obviously also exist.

This fact generates a further question, namely about the entropy solution's independence of the choice of the entropy functional $\tilde{S}$.

In order to answer this question, we cite the following theorem whose proof is more or less just a word-by-word copy of the proof of the Rankine–Hugoniot condition that takes (3.12) into account instead of (2.2):

Theorem 3.1

Assume that there is a pair $\tilde{S}$, $\tilde{F}$ of functions that belong to a weak solution V and fulfill condition (3.14); let the discontinuities of V form a curve Γ along which the Rankine–Hugoniot condition holds. Then it can be stated that V is a Lax entropy solution if and only if the inequality

$$\left[\tilde{S}\right] \hat{v} \leq \left[\tilde{F}\right] \tag{3.17}$$

holds, where $\hat{v}$ is the velocity that occurs in the Rankine–Hugoniot condition (2.15), i.e., the velocity with which the discontinuities move along the x-axis.

As in (2.15), $\left[\tilde{S}\right] := \tilde{S}(V_{\mathrm{L}}) - \tilde{S}(V_{\mathrm{R}})$ and $\left[\tilde{F}\right] := \tilde{F}(V_{\mathrm{L}}) - \tilde{F}(V_{\mathrm{R}})$ denote the heights of the jumps.

In the scalar case, and together with (2.15), (3.17) yields a particular form of the entropy condition equivalent to (3.12), namely

$$\frac{\left[\tilde{S}\right][f]}{[v]} \leq \left[\tilde{F}\right]\ , \tag{3.18}$$

i.e., using (3.13),

$$\frac{\tilde{S}(v_{\mathrm{L}}) - \tilde{S}(v_{\mathrm{R}})}{v_{\mathrm{L}} - v_{\mathrm{R}}}\left(f(v_{\mathrm{L}}) - f(v_{\mathrm{R}})\right) - \int_{v_{\mathrm{R}}}^{v_{\mathrm{L}}} \tilde{S}'(\alpha) f'(\alpha)\, \mathrm{d}\alpha \leq 0\ . \tag{3.19}$$

45) As has been assumed to be the case so far

This leads to the following theorem:

Theorem 3.2

If the entropy functional $\tilde{S}(v)$ is strictly convex,[45] i.e.,

$$\tilde{S}''(v) > 0\,, \quad \forall\, v \in \mathbb{R}\,,$$

and if the flux f is strictly convex, and also if discontinuities only occur along smooth curves Γ which do not intersect in the area under consideration, condition (3.19) only holds if and only if the jump relation

$$v_{\mathrm{L}} \geq v_{\mathrm{R}} \tag{3.20}$$

is respected along each of these curves Γ. In particular, this result shows that all entropy conditions created using strictly convex functionals $\tilde{S}$ and their entropy fluxes $\tilde{F}$ are equivalent, because these functionals do not play a role within Eq. (3.20).

Proof:

The function $\tilde{f}(v)$ defined for every fixed value of v_r by

$$\tilde{f}(v) := \begin{cases} \dfrac{\tilde{S}(v) - \tilde{S}(v_{\mathrm{R}})}{v - v_{\mathrm{R}}}(f(v) - f(v_{\mathrm{R}})) - \displaystyle\int_{v_{\mathrm{R}}}^{v} \tilde{S}'(\alpha) f'(\alpha)\, \mathrm{d}\alpha & \text{for } v \neq v_{\mathrm{R}} \\ 0 & \text{for } v = v_{\mathrm{R}} \end{cases}$$

is differentiable, and its derivative reads

$$\tilde{f}'(v) = \begin{cases} -\left\{\tilde{S}'(v) - \dfrac{\tilde{S}(v) - \tilde{S}(v_{\mathrm{R}})}{v - v_{\mathrm{R}}}\right\} \cdot \left\{f'(v) - \dfrac{f(v) - f(v_{\mathrm{R}})}{v - v_{\mathrm{R}}}\right\} & \text{for } v \neq v_{\mathrm{R}} \\ 0 & \text{for } v = v_{\mathrm{R}} \end{cases}\,.$$

The terms within the braces are positive for $v > v_{\mathrm{R}}$ and negative for $v < v_{\mathrm{R}}$ because $\tilde{S}$ as well as f are strictly convex. This leads to

$$\tilde{f}'(v) < 0 \qquad \text{for} \qquad v \neq v_{\mathrm{R}}$$

so that v_{R} is the only value for which $\tilde{f}$ vanishes, and $\tilde{f}$ decreases strictly monotonously for $v \neq v_{\mathrm{R}}$. Thus,

$$\tilde{f}(v) > 0 \text{ for } v < v_{\mathrm{R}}\,, \quad \tilde{f}(v) = 0 \text{ for } v = v_{\mathrm{R}}\,, \quad \tilde{f}(v) < 0 \text{ for } v > v_{\mathrm{R}}\,. \tag{3.21}$$

Equation (3.19) implies that $\tilde{f}(v_{\mathrm{L}}) \leq 0$, and because of (3.21) this is really equivalent to

$$v_{\mathrm{L}} \geq v_{\mathrm{R}}\,. \tag{3.22}$$

Remark

The same arguments, if applied to our traffic flow example with its strictly *concave flux* $f(\varrho)$ would similarly lead to the equivalence of the entropy condition to

$$\varrho_L \leq \varrho_R,^{46)} \tag{3.23}$$

and the interpretation of this result shows that car drivers endeavor to smooth a discontinuous situation as quickly as possible, i.e., to reach an equals sign,[47] or that they aim to achieve a nondecreasing development of the traffic density close to an actual discontinuity. In particular, car drivers do not stop when they notice a traffic jam ahead; they would rather drive to the end of the jam so that the density increases there. When a driver stops, the end of the jam (i.e., the discontinuity) passes over him/her, moving against the direction that the cars are traveling in.

The particular weak solution (2.11) of our traffic flow example obviously *does not* fulfill condition (3.23). In other words, (3.23) represents the driver's ride impulse and completes the rough Lighthill–Whitham traffic flow model.[48]

There is also a geometric characterization of the entropy solution, provided that the flux f is strictly convex:

The Rankine–Hugoniot condition (2.15), together with the mean value theorem, yields

$$\hat{v} = \frac{f(v_L) - f(v_R)}{v_L - v_R} = f'(v_R + \vartheta(v_L - v_R)) \, , \quad (0 < \vartheta < 1) \, ,$$

and the strict convexity of f together with (3.20) then leads to

$$f'(v_R) < \hat{v} < f'(v_L) \, . \tag{3.24}$$

Thus, if and only if the weak solution is the entropy solution, the slopes of the characteristics (1.17) belonging to this solution cause the characteristics to run into the shock curves Γ for increasing t instead of leaving them (cf. Fig. 3.2).

46) The strictly concave case can easily be transformed into a problem with a strictly convex flux.

47) The discontinuity at the front of a traffic jam is just such a situation.

48) R. Ansorge: Transpn. Res. **24B** (1990) 133–143

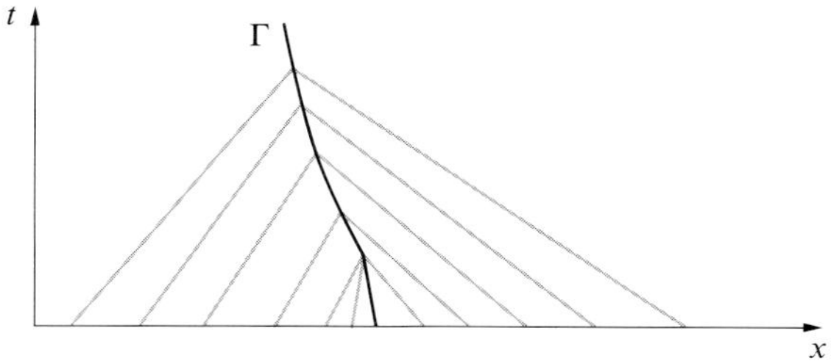

Fig. 3.2 Course of the characteristics for entropy solutions of scalar problems as soon as shocks appear.

As far back as 1957, Olga Oleinik published an entropy condition for weak solutions of scalar conservation laws.[49] It can be formulated as follows:

Definition

A weak solution v of the scalar problem (2.1) is called an entropy solution in the sense of Oleinik – or an Oleinik solution – if there is a constant $\mathcal{E} > 0$ such that

$$\frac{v(x+a,t) - v(x,t)}{a} \leq \frac{\mathcal{E}}{t}, \quad \forall a > 0, \quad \forall t > 0, \quad \forall x \in \mathbb{R}. \tag{3.25}$$

Assume that the discontinuities of a weak solution of a one-dimensional scalar conservation law form a curve Γ such that the Rankine–Hugoniot condition (2.15) holds along this curve; if then the point $(x, t) \in \Omega$ ($t > 0$, fixed) moves to Γ from the left, the point $(x + a, t), a > 0$, eventually becomes a point to the right of Γ such that (3.25) states

$$v(x+a,t) - v_L \leq \frac{\mathcal{E}}{t} a .$$

If a now tends to zero,

$$v_R - v_L \leq 0$$

follows.

Remark

Thus, an Oleinik solution is a Lax solution, too. If – hopefully – a uniqueness theorem for the Lax entropy solution is available, the Oleinik solution is also uniquely determined and equals the Lax solution.

This uniqueness problem for one-dimensional scalar conservation laws will be discussed in the next section, and the question concerning a suitable generalization of the definition of a Lax entropy solution to systems of conservation laws will then also be addressed.

49) O. Oleinik: Usp. Math. Nauk. (N.S.) **12** (1957) 3–73

Remark

Another approach to the idea of an entropy solution for scalar problems comes from the well-known uniqueness of the solution $v_\varepsilon(x, t)$ of the parabolic problem

$$\begin{aligned} \partial_t v + \partial_x f(v) &= \varepsilon \partial_{xx} v \,, \quad \varepsilon > 0 \\ v(x, 0) &= v_0(x) \,, \end{aligned}$$

if the right hand side of (2.1) is enriched – analogously to the linear problem (1.24) – by a diffusion term where the diffusion parameter ε tends to 0.[50]

3.3 Uniqueness of Entropy Solutions

In this section we are going to investigate the uniqueness of the Lax entropy solution. Lax himself gave the outline of the following theorem:

Theorem 3.3

Let $\tilde{v}$ and v be two Lax entropy solutions of the scalar form of the conservation law (2.1) which are piecewise continuous for every fixed $t \geq 0$, non-smooth only along certain curves Γ, and have initial functions $\tilde{v}_0$ and v_0, respectively. The flux f is assumed to be strictly convex. Then, the distance

$$\|\tilde{v}(\cdot, t) - v(\cdot, t)\|_{L_1(I)}$$

monotonously decreases with respect to t, where the L_1-norm is understood to be taken over an arbitrary but fixed interval I of the x-axis.[51]

Proof:

Decompose, for every fixed t, the interval I into intervals $(x_\nu, x_{\nu+1})$ $(\nu = 0, \pm 1, \pm 2, \ldots)$ in such way that $\tilde{v}(x, t) - v(x, t)$ has the sign $(-1)^\nu$ uniformly on $(x_\nu, x_{\nu+1})$. A decomposition of this type exists because $\tilde{v} - v$ is assumed to be piecewise continuous with respect to x. Evidently, x_ν depends on t: $x_\nu = x_\nu(t) \quad (\nu = 0, \pm 1, \pm 2, \ldots)$. Thus,

$$\|\tilde{v}(\cdot, t) - v(\cdot, t)\|_{L_1(I)} = \sum_\nu (-1)^\nu \int\limits_{x_\nu(t)}^{x_{\nu+1}(t)} \left[\tilde{v}(x, t) - v(x, t)\right] \, \mathrm{d}x \,. \tag{3.26}$$

50) N.N. Kuznetsov: USSR Comput. Math. Math. Phys. **16** (1976) 105–119

51) In particular, if then $\tilde{v}_0 = v_0$ in L_1, i.e., $\|\tilde{v}(\cdot, 0) - v(\cdot, 0)\|_{L_1(I)} = 0$ for every interval I, $\|\tilde{v}(\cdot, t) - v(\cdot, t)\|_{L_1(I)} = 0$ follows for all $t \geq 0$, i.e., $\tilde{v} = v$ in L_1 at every time level t, and this means uniqueness.

$\tilde{v}$ and v are piecewise smooth on $(x_\nu(t), x_{\nu+1}(t))$, and differentiation with respect to t yields

$$\frac{d}{dt}\|\tilde{v}(\cdot,t) - v(\cdot,t)\|_{L_1(I)} = \sum_\nu (-1)^\nu \left\{ \left[\tilde{v}(x_{\nu+1}(t),t) - v(x_{\nu+1}(t),t)\right] x'_{\nu+1}(t) \right.$$
$$- \left[\tilde{v}(x_\nu(t),t) - v(x_\nu(t),t)\right] x'_\nu(t)$$
$$\left. + \int_{x_\nu(t)}^{x_{\nu+1}(t)} \left[\partial_t\tilde{v} - \partial_t v(x,t)\right] dx \right\} .$$

Using (2.1),

$$\int_{x_\nu(t)}^{x_{\nu+1}(t)} \partial_t v(x,t)\, dx = -\int_{x_\nu(t)}^{x_{\nu+1}(t)} \partial_x f(v(x,t))\, dx$$
$$= -f(v(x_{\nu+1}(t),t)) + f(v(x_\nu(t),t))$$

follows; analogously for $\tilde{v}$.

Hence,

$$\frac{d}{dt}\|\tilde{v}(\cdot,t) - v(\cdot,t)\|_{L_1(I)} = \sum_\nu (-1)^\nu \left\{ \left[\tilde{v}(x_{\nu+,1}(t),t) - v(x_{\nu+1}(t),t)\right] x'{}_{\nu+1}(t) \right. \tag{3.27}$$
$$- \left[\tilde{v}(x_\nu(t),t) - v(x_\nu(t),t)\right] x'{}_\nu(t)$$
$$- \left[f(\tilde{v}(x_{\nu+1}(t),t)) - f(v(x_{\nu+1}(t),t))\right]$$
$$\left. + \left[f(\tilde{v}(x_\nu(t),t)) - f(v(x_\nu(t),t))\right] \right\} .$$

We have to distinguish between two different situations:

i) The function $\tilde{v}(\cdot,t_0) - v(\cdot,t_0)$ is continuous at $x_\nu(t_0)$.[52] This function then vanishes at $x_\nu(t_0)$, i.e.:

$$\tilde{v}(x_\nu(t_0),t_0) = v(x_\nu(t_0),t_0) . \tag{3.28}$$

ii) One of the functions $\tilde{v}(x,t_0)$, $v(x,t_0)$ is discontinuous at $x_\nu(t_0)$.

We therefore need to sum up the right hand side of (3.27) only for the particular numbers ν such that $(x_\nu(t),t)$ or $(x_{\nu+1}(t),t)$ lie on curves of discontinuities of $\tilde{v}$ or v.

52) For the arbitrary fixed value of t_0 under consideration

Let us first discuss the case where $(x_{\nu+1}(t), t)$ lies on a curve of discontinuities of $\tilde{v}$, whereas v is smooth at this position. The end-point $x_{\nu+1}$ of the interval $(x_\nu(t), x_{\nu+1}(t))$ must then be looked upon as $x_{\nu+1} - 0$, i.e., $\tilde{v}(x_{\nu+1}(t), t) = \tilde{v}_\ell$. If, for example, $\tilde{v} - v > 0$ on $(x_\nu, x_{\nu+1})$, and thus $\tilde{v} - v < 0$ on $(x_{\nu+1}, x_{\nu+2})$, then ν is even and

$$\tilde{v}_{\mathrm{L}} > v_{\mathrm{L}} = v_{\mathrm{R}} > \tilde{v}_{\mathrm{R}} \tag{3.29}$$

because $x_{\nu+1}$ on $(x_{\nu+1}, x_{\nu+2})$ has to be viewed as $x_{\nu+1} + 0$.

This leads to

$$\begin{aligned}(-1)^\nu \Big\{ &\left[\tilde{v}(x_{\nu+1}(t), t) - v(x_{\nu+1}(t), t)\right] x'_{\nu+1}(t) \\ &- \left[f(\tilde{v}(x_{\nu+1}(t), t)) - f(v(x_{\nu+1}(t), t))\right] \Big\} \\ = (\tilde{v}_{\mathrm{L}} - v_{\mathrm{L}}) &x'_{\nu+1}(t) - f(u_{\mathrm{L}}) + f(v_{\mathrm{L}}) \,.\end{aligned}$$

Because of the smoothness of v at $(x_{\nu+1}(t), t)$, $x = x_{\nu+1}(t)$ can only be a point on the shock curve of $\tilde{v}$. The Rankine–Hugoniot condition (2.15) therefore makes it possible to estimate the right hand side of the last equation by

$$\begin{aligned}(\tilde{v}_{\mathrm{L}} - v_{\mathrm{L}}) &\frac{f(\tilde{v}_{\mathrm{L}}) - f(\tilde{v}_{\mathrm{R}})}{\tilde{v}_{\mathrm{L}} - \tilde{v}_{\mathrm{R}}} - f(\tilde{v}_{\mathrm{L}}) + f(v_{\mathrm{L}}) \\ &= (\tilde{v}_{\mathrm{L}} - v_{\mathrm{L}}) \left[\frac{f(\tilde{v}_{\mathrm{L}}) - f(\tilde{v}_{\mathrm{R}})}{\tilde{v}_{\mathrm{L}} - \tilde{v}_{\mathrm{R}}} - \frac{f(\tilde{v}_{\mathrm{L}}) - f(v_{\mathrm{L}})}{\tilde{v}_{\mathrm{L}} - v_{\mathrm{L}}}\right] \le 0 \,,\end{aligned}$$

which takes into account the fact that (3.29) yields $\tilde{v}_{\mathrm{L}} - v_{\mathrm{L}} > 0$, but that

$$\frac{f(\tilde{v}_{\mathrm{L}}) - f(\tilde{v}_{\mathrm{R}})}{\tilde{v}_{\mathrm{L}} - \tilde{v}_{\mathrm{R}}} - \frac{f(\tilde{v}_{\mathrm{L}}) - f(v_{\mathrm{L}})}{\tilde{v}_{\mathrm{L}} - v_{\mathrm{L}}} \le 0 \,.$$

Here, the last inequality results from $\tilde{v}_{\mathrm{R}} < v_{\mathrm{L}} < \tilde{v}_{\mathrm{L}}$, which also follows from (3.29), as well as from the strict convexity of f (cf. Fig. 3.3).

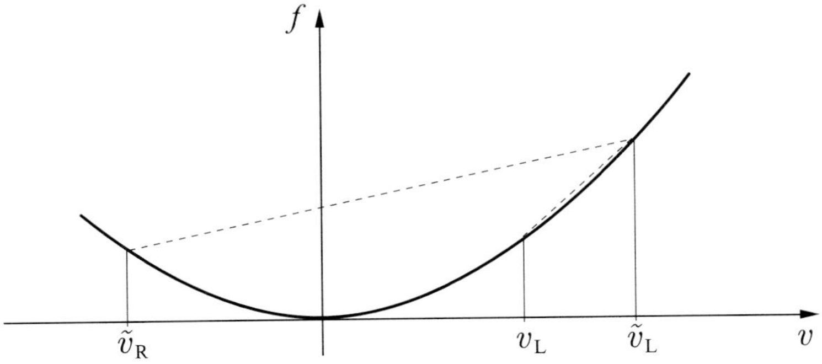

Fig. 3.3 Sketch relating to Lax' uniqueness theorem.

A similar result follows if $(x_{\nu+1}(t), t)$ is a point on a curve which is a set of discontinuities of $\tilde{v}$ as well as v; in this situation, the inequality

$$\tilde{v}_L > v_L > v_R > \tilde{v}_R$$

must be taken into account instead of (3.29), and it is to this part of the proof that the jump condition $v_L > v_R$ applies.

The same reflections work for the values of $\tilde{v}$ and v at the position $(x_\nu(t), t)$, namely the disappearance or non-positivity of the terms. Here, $x_\nu(t) = x_R$ must be observed in the case of a discontinuity.

The case $\tilde{v} < v$ on $(x_\nu, x_{\nu+1})$ can be treated in the same way. Thus,

$$\frac{d}{dt}\|\tilde{v}(\cdot, t) - v(\cdot, t)\|_{L_1(I)} \leq 0$$

follows.

As already emphasized, for systems of conservation laws with more than two equations, pairs $(\tilde{S}, \tilde{F})$ of an entropy functional $\tilde{S}$ and an entropy flux $\tilde{F}$ connected by (3.14) and suitable for distinguishing an entropy solution from other weak solutions do not necessarily exist.

We continue to treat only one-dimensional problems, but we try to formulate a suitable analog to (3.24) for problems where the convexity of the flux can also be suitably generalized.

Let $N \subset \mathbb{R}^m$ be a subset of the m-dimensional Euclidian space, and assume that system (2.1) is strictly hyperbolic for all $V = (v_1, v_2, \ldots, v_m)^T \in N$, i.e., the eigenvalues $\lambda_i(V)$ of the Jacobian $Jf(V)$ are real and different from each other for all $V \in N$. Moreover, let $f(V) \in C^2(N)$ and

$$\lambda_1(V) < \lambda_2(V) < \ldots < \lambda_m(V) . \tag{3.30}$$

The eigenvalues maintain this order for all $V \in N$; otherwise a vector $V \in N$ with

$$\lambda_i(V) = \lambda_k(V) \quad \text{for a pair} \quad (i, k) \quad \text{with} \quad i \neq k$$

would have to exist because of continuity reasons. However, this would contradict the strict hyperbolicity on N.

The eigenvectors $s_i(V)$ $(i = 1, 2, \ldots, m)$ belonging to the eigenvalues $\lambda_i(V)$ $(i = 1, 2, \ldots, m)$, respectively, are linearly independent.

Let $V(x, t)$ be a solution of (2.1) which is smooth on the domain $G \subset \Omega$ and has

$$V(x, t) \in N , \qquad \forall\, (x, t) \in G .$$

The i-characteristics $x = x_{(i)}(t)$ corresponding to this solution (cf. (1.18)) are uniquely determined from the Picard–Lindeløf theorem if the functions $\lambda_i(V(x, t))$ fulfill a Lipschitz condition with respect to x $(i = 1, \ldots, m)$ and provided initial states $x_{(i)}(t_0)$ with $(x_{(i)}(t_0), t_0) \in G$ are prescribed. Of course, existence and uniqueness theorems with weaker assumptions are also available, e.g., the existence theorem of Peano combined with the uniqueness condition of Nagumo.

Then, along an i-characteristic,

$$\frac{\mathrm{d}}{\mathrm{d}t} V(x_{(i)}(t), t) = -\left[Jf(V(x_{(i)}(t), t)) - \lambda_i(V(x_{(i)}(t), t)) \cdot \boldsymbol{I} \right] \partial_x V(x_{(i)}(t), t) , \qquad (3.31)$$

where $\boldsymbol{I}$ in this context means the $m \times m$ unit matrix.

In the case of $m = 1$, the right hand side of (3.31) vanishes, as observed earlier, so that $V = v$ is constant along a characteristic which therefore becomes a straight line.

Does an analog exist for $m > 1$? In other words, do a weak solution V and a set of i-characteristics $x_{(i)}$ belonging to this solution exist such that the solution is constant along these characteristics, which therefore become straight lines?

In the case of $\partial_x V(x_{(i)}(t), t) \neq 0$, a positive answer requires that the following relation is valid:

$$\partial_x V(x_{(i)}(t), t) = s_i(V(x_{(i)}(t), t)) \quad \forall\, t > 0 \, .$$ [53]

Solutions of this type can actually be specified. If we introduce so-called *centered waves* $V = V^*$, which are centered at a position $P_0 = (x_0, t_0) \in G$, i.e., waves of the form

$$V(x, t) = V^* \left(\frac{x - x_0}{t - t_0} \right) \qquad (3.32)$$

with

$$V^*(\xi) \in N \cap \left[C^1(\mathbb{R}) \right]^m \quad \forall\, P = (x, t) \in G \, , \quad \xi := \frac{x - x_0}{t - t_0} \, .$$

These centered waves solve (2.1) if they fulfill

$$\partial_t V + \partial_x f(V) = \left(\partial_x \xi J f(V^*) + \partial_t \xi I \right) V^{*\prime}(\xi) = 0$$

which can also be described as

$$(Jf(V^*(\xi)) - \xi I) \, V^{*\prime}(\xi) = 0 \, . \qquad (3.33)$$

Hence, for one $i \in (1, 2, \ldots, m)$, the pair of relations

$$\begin{aligned} V^{*\prime}(\xi) &= s_i\left(V^*(\xi)\right) , \\ \xi &= \lambda_i\left(V^*(\xi)\right) \qquad \text{(for all possible values of } \xi) \end{aligned} \qquad (3.34)$$

must hold.

In this case, differentiation of the second equation of (3.34) with respect to ξ together with the first equation yields (cf. footnote 51)

$$1 = \langle \nabla_V \lambda_i(V), s_i(V) \rangle \, . \qquad (3.35)$$

Definition

If $\langle \nabla_V \lambda_i(V), s_i(V) \rangle \neq 0$ holds for all $V \in N$ with a set $N \subset \mathbb{R}^m$, the *i-th characteristic field* $\left(\lambda_i(V), s_i(V)\right)$ is termed *genuinely nonlinear* on N.

53) Here, a suitable norm of the vectorfield V is chosen.

Remark

Because the scalar case implies $\lambda(V) = \lambda(v) = f'(v)$, which yields

$$\nabla_V \lambda(V) = f''(v) \,,$$

(3.35) can only be fulfilled if $f''(v) \neq 0$, and $s(v)$ in this case is a nonzero scalar factor $s(v)$. It can be normalized by $s(v) = \frac{1}{|f''(v)|}$, which makes (3.35) equivalent to $f''(v) > 0$, i.e., to strict convexity of the flux (or, if we normalize by $\frac{-1}{|f''(v)|}$, to strict concavity).

We are therefore going to consider (3.35) as the transition announced earlier of the strict convexity of the flux as defined for scalar problems to the i-th characteristic field in the case of $m > 1$.

Let us now assume that N is a simply connected domain in $\mathbb{R}^m$. Moreover, let $V^0 \in N\left(V^0 \neq 0\right)$ be a given *state* and $\xi_0 := \lambda_i\left(V^0\right)$. If ξ_0 is used as starting point with an initial condition

$$V^*(\xi_0) = V^0 \,, \tag{3.36}$$

the second equation in (3.34) will obviously be fulfilled at this point ξ_0.

The first equation of (3.34) represents a system of ordinary differential equations where (3.36) is used as the initial condition. We know that there is a unique solution for this system in a certain neighborhood of (ξ_0, V^0) provided that the assumptions of a standard theorem like the Picard–Lindeløf theorem are fulfilled. This solution runs with respect to V in N; let's say for $\xi_0 - a \leq \xi \leq \xi_0 + a$ with sufficiently small $a > 0$. Also, (3.35) then holds for $V = V^*(\xi)$ because of the assumptions.

Since the first equation of (3.34) was also fulfilled, (3.33) also holds. Integration of (3.33) together with (3.36) then yields the fact that the second equation of (3.34) is also fulfilled, at least for the values of ξ considered here and if a can be diminished somewhat if necessary.

We conclude that there is a nonvanishing smooth solution V, namely a wave centered at P_0, which belongs to a given nonvanishing smooth initial state V^0 and to a particular integer i, and leads to an i-th characteristic field which is genuinely nonlinear on N:

$$V(x,t) = V^*\left(\frac{x-x_0}{t-t_0}\right) \,, \qquad V^*\left(\lambda_i(V^0)\right) = V^0 \tag{3.37}$$

for all $(x, t) \in \Omega$ with

$$\lambda_i(V^0) - a \leq \frac{x-x_0}{t-t_0} \leq \lambda_i(V^0) + a \quad \text{where} \quad a > 0 \quad \text{is sufficiently small} \,. \tag{3.38}$$

Along each of the straight lines

$$\frac{x-x_0}{t-t_0} = c = \text{const}$$

with

$$\lambda_i\left(V^0\right) - a \leq c \leq \lambda_i\left(V^0\right) + a$$

forming a fan, the solution

$$V^*\left(\frac{x - x_0}{t - t_0}\right) = V^*(c)$$

is constant.

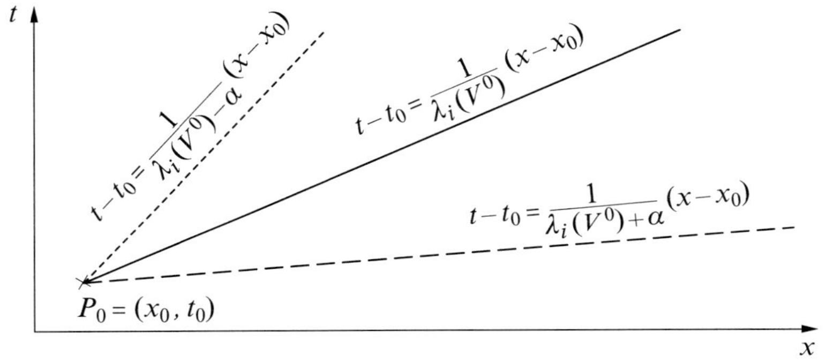

Fig. 3.4 Fan of i-characteristics of centered waves.

Thus, along these straight lines given by

$$x_{(i)}(t) = x_0 + c(t - t_0) \ ,$$

the relation

$$\frac{d}{dt} V(x_{(i)}(t), t) = \frac{d}{dt} V^*(c) = 0 \tag{3.39}$$

holds. However, because $V^{*\prime}(c)$ is an eigenvector belonging to $\lambda_i(V^*(c))$, this also applies to

$$\partial_x V(x_{(i)}(t), t) = \frac{1}{t - t_0} V^{*\prime}(c) \ .$$

It follows from (1.18) and (3.33) that the straight lines that generate the fan and belong to this particular integer i are the i-characteristics formed using solution (3.32) through the point P_0, and the centered waves constructed so far are constant along these lines.

Let us now assume that V^1 is a state in the neighborhood of V^0. Then, by continuity arguments, $\lambda_i\left(V^1\right)$ lies in a neighborhood of $\lambda_i\left(V^0\right)$ such that

$$\left|\lambda_i\left(V^1\right) - \lambda_i\left(V^0\right)\right| \leq a$$

can be fulfilled for all states V^1 sufficiently close to V^0.

In particular, if

$$\lambda_i(V^0) < \lambda_i(V^1) \le \lambda_i(V^0) + a \text{ ,}^{54)}$$

we obtain the result

$$V^*\left(\frac{x - x_0}{t - t_0}\right) = V^0$$

along the straight lines

$$t - t_0 = \frac{1}{\lambda_i\left(V^0\right)}(x - x_0) \text{ ,}$$

so the solution $V^*(\xi)$ is constant along the lines and equals V^0 if this is its value at ξ_0. The same arguments show that the solution equals V^1 along the straight lines

$$t - t_0 = \frac{1}{\lambda_i\left(V^1\right)}(x - x_0)$$

if this is its value at ξ_0.

In other words, in the case of $\lambda_i\left(V^0\right) < \lambda_i\left(V^1\right)$, the state V^0 can be *smoothly connected from the right* with the state V^1 by a centered i-rarefaction wave.[55)]

Remark

In the scalar case, and if f is strictly convex, $\lambda_i\left(V^0\right) < \lambda_i\left(V^1\right)$ means $f'\left(v^0\right) < f'\left(v^1\right)$, and so

$$v^0 < v^1 \text{ .}$$

In other words, if t is fixed, the states decrease for decreasing x, i.e., from the right to the left (rarefaction).[56)] Indeed, if the entropy condition (3.20) is fulfilled, this rarefaction can only be caused by a continuous process, because $v_L = v_R$.

Assume now that system (2.1) has discontinuous solutions as well as or instead of smooth rarefaction waves, and let us assume that these discontinuities form certain curves Γ along which the Rankine–Hugoniot condition is satisfied.

We again require conditions that are easy to apply in order to *uniquely* characterize the physically relevant entropy solution.[57)]

Here, we restrict ourselves to problems whose characteristic fields are genuinely nonlinear on $N = \mathbb{R}^m$.[58)]

54) In the case of $\lambda_i(V^1) < \lambda_i(V^0)$, swap the roles of V^1 and V^0.

55) See also the definition that follows Eq. (3.43)

56) If the flux f is a concave function, like the flux in the traffic flow model, the word "left" should be replaced by "right" and "decrease" by "increase."

57) Recall that condition (3.12) cannot necessarily be applied when there are more than two equations.

58) This is analogous to scalar problems, i.e., to $m = 1$ where only convex fluxes are taken into account.

A weak solution will be called an entropy solution if it respects a condition of the following type:

- If initial values are prescribed along a non-characteristic curve, the condition will only be fulfilled by one weak solution
- The condition coincides with (3.24) for scalar problems

If $[V] = V_L - V_R$ are the jumps along a curve Γ of discontinuities of a weak solution moving along the x-axis at a speed $\hat{v}$, let k be the particular integer with

$$\lambda_k(V_R) < \hat{v} < \lambda_{k+1}(V_R)\,, \quad k \in \{1, 2, \ldots, m-1\}\,, \tag{3.40}$$

or $\hat{v} < \lambda_1(V_R)$ or $\hat{v} > \lambda_m(V_R)$.[59]

Only the values of V_0 along the particular characteristics through P_0 which show the steeper ascents

$$\frac{1}{\lambda_k(V_0)}\,, \quad \frac{1}{\lambda_{k-1}(V_0)}\,, \quad \ldots\,, \quad \frac{1}{\lambda_1(V_0)} \quad \text{(cf. Fig. 3.5)}$$

are available in order to determine the m components of V_R at $P_0 \in \Gamma$ from the given initial values.

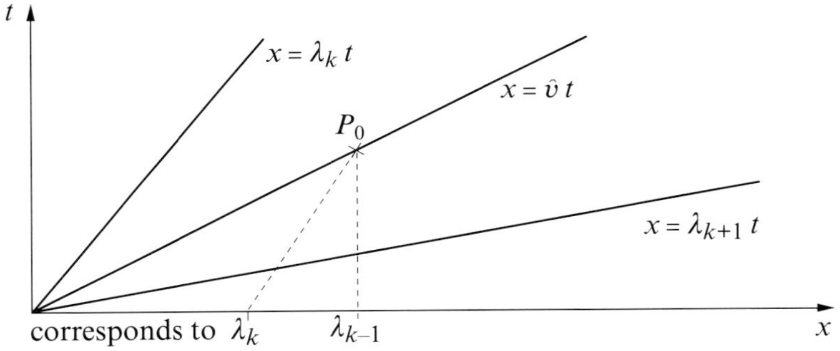

Fig. 3.5 Sketch concerning the completeness of the information needed in order to ensure uniqueness.[60]

If, for example, a system of decoupled equations is under consideration,[61] one and only one of the m components of $V_R(P_0)$ can be determined along each of these k steeper characteristics. Hence, $m - k$ pieces of information are missing.

If, at the same time,

$$\lambda_j(V_L) < \hat{v} < \lambda_{j+1}(V_L)\,,$$

holds, corresponding arguments lead to the result that there are only $m - j$ pieces of information available for the computation of the m components of V_L. Hence, j pieces of information are missing.

59) See (3.30)

60) The particular choice of $0 \in \Gamma$ in Fig. 3.5 does not lead to a loss of generality.

61) See the linear situation (1.19)

Thus, $m - k + j$ pieces of information are missing for the unique determination of the two quantities V_L and V_R. However, m additional pieces of information concerning the relations between V_L, V_R and $\hat{v}$ follow from the Rankine–Hugoniot condition (2.15). One of them must be used for the elimination of $\hat{v}$. Thus, the real number of pieces of missing information is only $1-k+j$. In other words, *uniqueness* can only be expected for

$$k = j + 1 ,$$

i.e., if the relations

$$\lambda_k(V_R) < \hat{v} < \lambda_{k+1}(V_R) , \quad \lambda_{k-1}(V_L) < \hat{v} < \lambda_k(V_L) \tag{3.41}$$

hold simultaneously.

Although this result was only illustrated by an example, we accept that it provides a hint about a general definition of an entropy solution:

Definition

V is called an *entropy solution* if there is an index k for which (3.41) holds. A discontinuity of this type is then called a *k-shock* and the inequalities (3.41) are called *entropy inequalities* or *(Lax) shock conditions.*

Equation (3.41) includes

$$\lambda_k(V_R) < \hat{v} < \lambda_k(V_L) ,$$

so that (3.41) really coincides with (3.24) as far as scalar problems are concerned. Thus, both of the properties claimed earlier with respect to the formulation of an entropy condition for weak solutions of systems of conservation laws are respected.

There are some connections between an i-characteristic field and the so-called i-Riemann invariant:

Definition

A scalar function $w = w(V)$ defined for all $V \in N \subset \mathbb{R}^m$ is called an *i-Riemann invariant* on N if

$$\langle \nabla_V w(V), s_i(V) \rangle = 0 \quad \forall\, V \in N . \tag{3.42}$$

Remark

Thus, the property that an i-characteristic field is genuinely nonlinear coincides with the property that $\lambda_i(V)$ is not an i-Riemann invariant.

If the graph of a centered i-rarefaction wave V is situated in N, its i-Riemann invariants are constant. This follows from (3.34):

$$\partial_x w\left(V(x, t)\right) = \langle \nabla_V w(V), s_i(V) \rangle_{V=V^*(x,t)} = 0 \tag{3.43}$$

and similarly $\partial_t w\left(V(x, t)\right) = 0$.

Definition

If a solution V of system (2.1) belonging to any initial function V_0 is smooth on a region $G \subset \Omega$, and if all the i-Riemann invariants $w(V(x,t))$ are constant on G, V is called an *i-simple wave* or an *i-rarefaction wave* on G.

3.4 Kruzkov's Ansatz

The approach used by S.N. Kruzkov[62] to introduce the idea of a weak solution of problem (2.1) seems to differ from that of Lax. In order to understand the situation, it is sufficient to restrict the presentation of Kruzkov's ansatz to one-dimensional scalar problems.

Kruzkov tries to find his weak solutions in the space $L_1^{\text{loc}}(\Omega)$, as already done in the Lax definition of a weak solution. Moreover, the test functions used in (2.2) by Lax also play a crucial role in Kruzkov's definition, with the one restriction that these functions now vanish along the whole boundary of their compact support, i.e., also along parts of the boundary belonging to the x-axis: $\Phi(x,0) \equiv 0$. This leads to the need to pay attention to the initial function in a different way than by the second integral in Eq. (2.2).

Kruzkov calls the function $v \in L_1^{\text{loc}}(\Omega)$ a weak solution of the problem

$$\partial_t v + \partial_x f(v) = 0\,, \quad v(x,0) = v_0(x)$$

if the following two conditions are fulfilled:

$$\int_\Omega \Big\{ \partial_t \Phi(x,t)\,|v(x,t) - c| + \partial_x \Phi(x,t) \\ \operatorname{sgn}\,(v(x,t) - c)\,[f(v(x,t)) - f(c)] \Big\}\, \mathrm{d}(x,t) \geq 0 \tag{3.44}$$

$$\forall\, \Phi \in C_0^1(\Omega) \quad \text{with} \quad \Phi \geq 0 \quad \text{and with} \quad \Phi(x,0) \equiv 0\,, \quad \forall\, c \in \mathbb{R}\,,$$

and

$$\lim_{t \to 0} \int_{-R}^{R} |v(x,t) - v_0(x)|\, \mathrm{d}x = 0 \qquad \forall\, R \in \mathbb{R} \quad (t \in [0,T] \backslash \tilde{\varepsilon}) \tag{3.45}$$

for $T > 0$ and for a set $\tilde{\varepsilon}$ of measure zero so that $v(x,t)$ is well defined almost anywhere as a function of x for any fixed $t \in [0,T] \backslash \tilde{\varepsilon}$.

Kruzkov was able to show that there is a unique weak solution in $L_1^{\text{loc}}(\Omega)$ in the sense of (3.44), (3.45) provided that the flux f is smooth and strictly convex. Smoothness and strict convexity of the flux are assumptions that were also made by Oleinik and Lax.

62) S.N. Kruzkov: Soviet Math. Dokl. **10** (1969) 785–788

Hence, in Kruzkov's definition, the characterization of an entropy solution does not consist of an Eq. (2.2) to be solved for all test functions together with an additional entropy condition (3.12), but simply an inequality.[63]

One can say that Kruzkov's definition already includes an entropy condition.

Indeed, as far as the scalar problem is concerned, the left side of (3.44) corresponds with (3.12) if $\Phi(x,0) = 0$ is taken into account, and if for every particular constant $c \in \mathbb{R}$ the entropy functional $\tilde{S}(v)$ is identified with $|v-c|$ and the entropy flux $\tilde{F}(v)$ with $\operatorname{sgn}(v-c)\,[f(v)-f(c)]$.

If $\tilde{S}$ and $\tilde{F}$ are chosen in this way, the connection between these two functions demanded by (3.13) is also kept piecewise, namely for all values of v with the exception of $v = c$. Moreover, the particular functional $\tilde{S}$ is also convex – as entropy functionals are expected to be – but unfortunately no longer strictly convex.

On the other hand, there is a somewhat stronger demand than formulated by Lax, namely that the validity of (3.44) is ensured for not only all nonnegative test functions Φ but also all $c \in \mathbb{R}$. Thus, (3.45) must now hold for all *test elements*

$$\hat{\Phi} \in \left\{(\Phi, c)\,|\,\Phi \in C_0^1(\Omega),\, \Phi \geq 0,\, \Phi(x,0) = 0;\, c \in \mathbb{R}\right\} . \tag{3.46}$$

It can be shown by more than one chain of evidence that Kruzkov's weak solution coincides with the Lax–Oleinik entropy solution provided that the flux is smooth and strictly convex. One of these chains follows from numerical aspects. There are finite difference approximations whose numerical solutions converge for decreasing step sizes to the Lax–Oleinik entropy solution, but also (as can be proved by a slightly different convergence argument) to the Kruzkov solution.[64]

It should be noted that problems can arise from fields of applications where the assumptions of smoothness and strict convexity of the flux are violated. However, Kruzkov, together with Panov, showed that a unique solution for the problem associated with (3.44) and (3.45) does exist, even if f is merely continuous.[65] Thus, in the scalar situation, the Kruzkov concept of an entropy solution is more far-reaching.

However, one has to pay for the renunciation of convexity and smoothness. It can happen that one of the properties of the entropy solution v_E that is important from the point of view of numerical methods gets lost; namely, that an initial function v_0 with compact support leads to a compact support of $v_E(\cdot, t)$ (for every fixed $t > 0$), as is the case if f is smooth and convex. A test example of this loss is given by

$$\begin{aligned} f(v) &= \frac{|v|^\alpha}{\alpha} \quad \text{with} \quad 0 < \alpha < 1 , \\ v_0(x) &= \begin{cases} 0 & \text{for} \quad x < -1 \\ 1 & \text{for} \quad -1 \leq x \leq 0 \\ 0 & \text{for} \quad x > 0 \end{cases} . \end{aligned} \tag{3.47}$$

63) As well as a condition that respects the initial condition in a suitable way

64) A non-numerical argument concerning the connection between both definitions can be found in the paper mentioned in footnote 2.

65) S.N. Kruzkov, E.Y. Panov: Soviet Math. Dokl. **42** (1991) 316–321

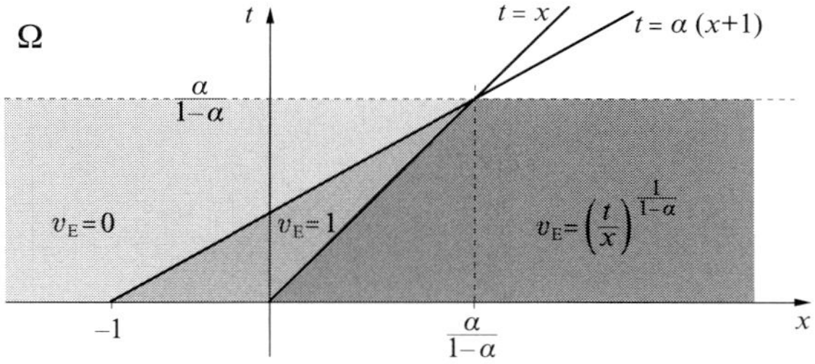

Fig. 3.6 Kruzkov–Panov solution (3.48) for $0 \le t \le \dfrac{\alpha}{1-\alpha}$.

The solution reads

$$v_E(x,t) = \begin{cases} 0 & \text{for} \quad t > \alpha(x+1) \\ 1 & \text{for} \quad x < t \le \alpha(x+1) \\ \left(\dfrac{t}{x}\right)^{\frac{1}{1-\alpha}} & \text{for} \quad t \le x \end{cases} \tag{3.48}$$

(cf. Fig. 3.6).

Whereas the smooth and strictly convex case behaves hyperbolically, a merely continuous flux can lead to a problem with more parabolic character. Numerical procedures for problems of this particular type were recently developed under suitable conditions for the first time by M. Breuss.[66] Here, the Kruzkov–Panov example (3.47) already shows that the important CFL condition (cf. Sections 6.1, 7.3) cannot be fulfilled when explicit finite difference methods (cf. Sections 4.1, 6.3, 7.2) are used. Thus, implicit methods must be taken into account, and this can actually be done successfully.

Examples of mathematical models of real-world problems in fluid mechanics, where standard assumptions like strict hyperbolicity, strict convexity or smoothness of flux are not necessarily fulfilled, occur in the theory of flows through porous media, e.g., in relation to oil production.

The scalar so-called nonstandard *Buckley–Leverett equation*

$$\partial_t v + \partial_x \left(\frac{v^2}{v^2 + a(1-v)^2} \right) = 0 \,, \quad a \in \mathbb{R}$$

models the displacement of an oleic phase by an aqueous phase in a porous medium. In this situation, the flux is neither convex nor concave.

Tveito and Winther[67] considered the following system theoretically with respect to stability:

$$\begin{aligned} \partial_t s + \partial_x f(s,c) &= 0 \\ \partial_t \left(sc + a(c)\right) + \partial_x \left(cf(s,c)\right) &= 0 \,. \end{aligned}$$

66) M. Breuss: *Numerik von Erhaltungsgleichungen in Nicht-Standard-Situationen.* Ph.D. Thesis, Dept. of Mathematics, University of Hamburg 2001

67) In: B. Engquist, B. Gustafsson (eds.): *Proceedings of the Third International Conference on Hyperbolic Problems.* Uppsala: Chartwell-Bratt 1991, pp. 888–898

This system also models the displacement of oil by water in a porous medium. However, in this case, the water is thickened by dissolved polymer in order to increase its viscosity. The difficulty of this example arises from the fact that the eigenvalues of the Jacobian are real but can coincide, so the system is hyperbolic but not necessarily strictly hyperbolic.

4
The Riemann Problem

4.1
Numerical Importance of the Riemann Problem

When we treated the traffic jam dissolution example in Section 2.2 we met the first time with a Riemann problem. It consisted of a conservation law of type (2.1) together with an initial function that was piecewise constant with certain jumps at isolated positions.

One tool that can be used to solve differential equations numerically is the so-called *finite difference method (FDM)*, where one tries to approximate the unknown values of the exact solution at isolated points (see also Chapters 6 and 7).

Particularly in the case of problem (2.1), we would like numerical values $V_\nu^{(n)}$ at isolated positions (x_ν, t_n) on the upper half of plane Ω that are expected to be good approximations to the values $V_E(x_\nu, t_n)$ of the exact entropy solution at these points.

One particular approach to constructing FDMs involves, in a first step, covering the positive time axis with a not necessarily equidistant *grid* $\{t_n|(n = 0, 1, \ldots)\}$ of *time step sizes* $\Delta t_n := t_{n+1} - t_n$ and fixing isolated grid points x_ν $(\nu = 0, \pm1, \pm2, \ldots)$ along the x-axis. Thus, by generating the Cartesian product of these two sets, Ω will be covered with a net of grid points parallel to the axes. Also, the step sizes $h_\nu := x_{\nu+1} - x_\nu$ do not necessarily need to be equidistant (cf. Fig. 4.1).

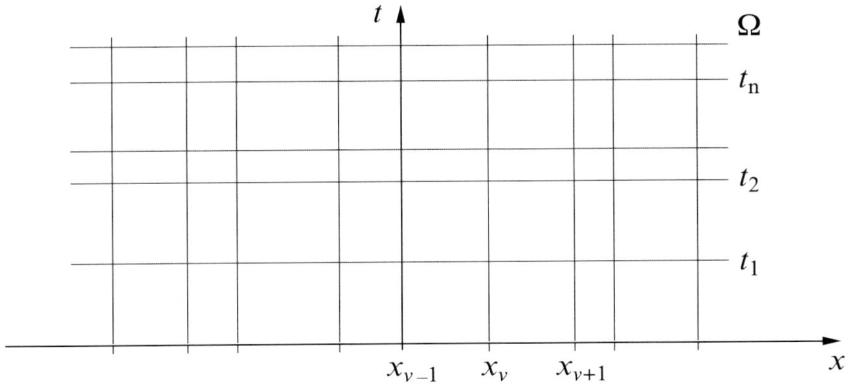

Fig. 4.1 Grid of an FDM.

Mathematical Models of Fluid Dynamics. R. Ansorge and T. Sonar

ISBN: 978-3-527-40774-3

If the values $V_\nu^{(n+1)}$ at time step $t = t_{n+1}$ are only computed based on knowledge of the values $V_\nu^{(n)}$ that were computed at the previous time step, we call the scheme a *one-step method*. In contrast, *multistep methods* include approximate values for earlier steps, but we are going to restrict the present survey to one-step methods.

Let $t_0 = 0$ be the initial level, with approximate values

$$V_\nu^{(0)} := \frac{2}{x_{\nu+1} - x_{\nu-1}} \int\limits_{\frac{1}{2}(x_{\nu-1}+x_\nu)}^{\frac{1}{2}(x_\nu+x_{\nu+1})} V_0(x)\,\mathrm{d}x \quad (\nu = 0, \pm 1, \pm 2, \ldots) . \tag{4.1}$$

It seems reasonable to describe the difference between the approximate solution and the unknown exact solution in terms of the topology of the space in which the original problem is based, which is the space $L_1^{\mathrm{loc}}(\Omega)$ in this particular case. This difference or *truncation error* should be estimated using a suitable norm. Hopefully, these errors will decrease for decreasing step sizes and will tend to zero sufficiently fast if the step sizes tend to zero. If this holds, the method is said to be *convergent*.

Obviously, a direct comparison of the approximate solution with the exact solution in terms of the toplology of the original space can happen if the discrete values $\left\{V_\nu^{(n)}\right\}$ of the approximate solution defined at isolated points can be appropriately extended to the intergrid points, e.g., by

$$\begin{aligned} &\hat{V}(x,t) := V_\nu^{(n)} \quad \text{for} \quad \frac{x_{\nu-1}+x_\nu}{2} < x \le \frac{x_\nu + x_{\nu+1}}{2}\,, \qquad t_n \le t < t_{n+1} \\ &(\nu = 0, \pm 1, \pm 2, \ldots; n = 0, 1, 2, \ldots) . \end{aligned} \tag{4.2}$$

Using this special extension or *reconstruction*, the discrete function now becomes a function $\hat{V}$ that is constant on each of the described rectangles. Hence, $\hat{V}$ is a step function on Ω and, thus, an element of $L_1^{\mathrm{loc}}(\Omega)$.

If we restrict $\hat{V}$ to the time step $t = t_n$, this restricted function $\hat{V}(x, t_n)$ of the space variable x is a step function, too, and from this, the one-step method together with the reconstruction now generate $\hat{V}(x, t_{n+1})$.

If $\hat{V}(x, t_n)$ is used as an initial function for the original problem in order to generate a weak solution of the differential equation in (2.1) for the region $t \ge t_n$, a Riemann problem arises. Its exact solution, if restricted to the time step $t = t_{n+1}$, would lead to values at the grid points at this step that could be considered approximations to the weak solution of the original problem at these new grid points. This idea was first used by S.K. Godunov,[68] and has since been used by various other authors as the basis for further numerical schemes. This enabled effective approximate *Riemann solvers* to be developed, including local linearizations.

We are therefore now going to study the Riemann problem, at least for linear problems with constant coefficients.

68) S.K. Godunov: Mat. Sb. **47** (1959) 357–393

4.2
The Riemann Problem for Linear Systems

Recall problem (1.19), i.e.,

$$\partial_t V + A\partial_x V = 0 \ ,$$

with a constant (m,m)-matrix A, and now prescribe the initial values

$$V_0(x) = \begin{cases} V_\ell & \text{for} \quad x < 0 \\ V_r & \text{for} \quad x > 0 \end{cases} \tag{4.3}$$

where the constant vectors V_ℓ and V_r are different from each other. Using the origin as the position of the jump causes no loss of generality.

Again assume the problem to be strictly hyperbolic, so that the eigenvalues λ_i of the matrix A are real and different from each other. The eigenvectors belonging to these eigenvalues are then obviously real too, and linearly independent. Thus, they can be used as a basis in $\mathbb{R}^m$. In areas where the solution is smooth, (1.21) and (1.22) yield:

$$V(x,t) = \sum_{i=1}^{m} \hat{v}_{i_0}(x - \lambda_i t) s_i \tag{4.4}$$

with

$$\hat{V}_0(x) = (\hat{v}_{1_0}(x), \hat{v}_{2_0}(x), \ldots, \hat{v}_{m_0}(x))^T = S^{-1} V_0(x) \ .$$

In particular, this leads to

$$\hat{V}_{0_\ell}(x) =: (w_1, w_2, \ldots, w_m)^T = S^{-1} V_\ell \quad \text{for} \quad x < 0$$

and analogously to

$$\hat{V}_{0_r}(x) =: (\tilde{w}_1, \tilde{w}_2, \ldots, \tilde{w}_m)^T = S^{-1} V_r \quad \text{for} \quad x > 0$$

with certain constants w_i and $\tilde{w}_i$.

If the eigenvalues are ordered as in (3.30),

$$V(x,t) = \sum_{i=1}^{k_0} \tilde{w}_i s_i + \sum_{i=k_0+1}^{m} w_i s_i \tag{4.5}$$

follows from (4.4), provided that for a given position (x,t) the first k_0 eigenvalues are the particular ones for which $x - \lambda_i t > 0$ $(i = 1, 2, \ldots, k_0)$, whereas $x - \lambda_i t < 0$ $(i = k_0 + 1, k_0 + 2, \ldots, m)$ holds for the other $m - k_0$ eigenvalues.

Let us now look at the i-characteristics $x = \lambda_i t$ arising from the jump position, and allow $t_0 > 0$ to be fixed. Let us move for this fixed t_0 from the left to the right along a line parallel to the x-axis. This line crosses the k_0-characteristic at the position $x_0 = \lambda_{k_0} t_0$ (cf. Fig. 4.2). As soon as this happens, the coefficient $\tilde{w}_{k_0}$ in Eq. (4.4) must be replaced by w_{k_0}.

The solution V described by (4.5) then obviously jumps. The height of this jump is

$$[V]_{k_0}(x_0, t_0) = (w_{k_0} - \tilde{w}_{k_0}) s_{k_0} \tag{4.6}$$

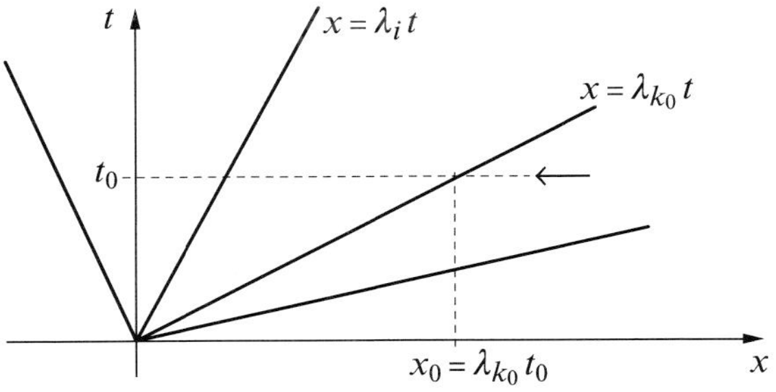

Fig. 4.2 How to solve the linear Riemann problem.

and leads to

$$A[V]_{k_0}(x_0, t_0) = (w_{k_0} - \tilde{w}_{k_0})\lambda_{k_0} s_{k_0} = \lambda_{k_0}[V]_{k_0}(x_0, t_0) \ . \tag{4.7}$$

We already know that discontinuities move along characteristics in linear problems. This makes the k_0-characteristic line through the jump position a k_0-shock Γ_{k_0} if $w_{k_0} \neq \tilde{w}_{k_0}$, meaning that (4.7) simply becomes the Rankine–Hugoniot condition (2.15) applied to the special linear situation.

Equations (4.5) and (4.7) yield the result that the solution to our Riemann problem for $t > 0$ can also be formulated as

$$V(x, t) = V_L - \sum_{\substack{i \\ \lambda_i < \frac{x}{t}}} (w_i - \tilde{w}_i) s_i \tag{4.8}$$

or as

$$V(x, t) = V_R + \sum_{\substack{i \\ \lambda_i > \frac{x}{t}}} (w_i - \tilde{w}_i) s_i \tag{4.9}$$

(cf. Fig. 4.3).

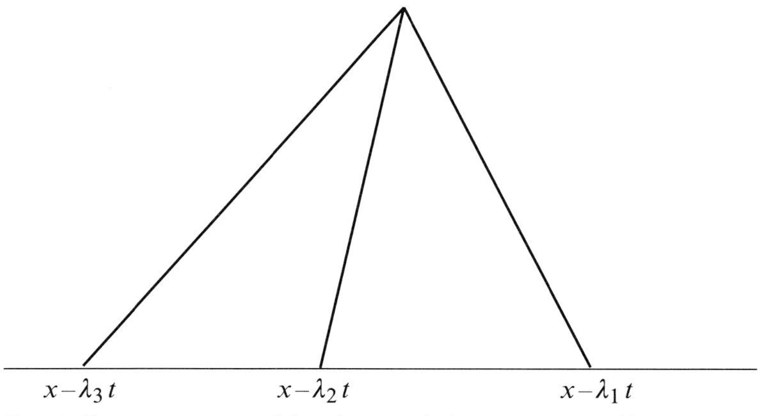

Fig. 4.3 Characteristic cones of the solution to the linear Riemann problem.

4.3 The Aw–Rascle Traffic Flow Model

The Riemann problem plays a significant role in a continuous traffic flow model which is more realistic than the Lighthill–Whitham model used in Chapter 2 to demonstrate the loss of uniqueness that can occur for weak solutions.

This more realistic mathematical description of traffic flow was created by A. Aw and M. Rascle.[69] The Lighthill–Whitham model does not respect the fact that a car – in contrast to gas particles – is only stimulated by the traffic flow situation in front of it, whereas a gas particle is also influenced by the situation behind it.

Taking this important difference into account, but retaining the idea of a continuous model, Aw and Rascle developed a model that leads to a system of two equations, namely to

$$\begin{aligned} \partial_t \varrho + \partial_x(\varrho v) &= 0 \\ \partial_t\big(v + p(\varrho)\big) + v\partial_x\big(v + p(\varrho)\big) &= 0\,, \end{aligned} \tag{4.10}$$

where $p(\varrho)$ is an increasing function with $p(\varrho) \sim \varrho^\gamma$ near $\varrho = 0$ $(\gamma > 0)$, as well as for all ϱ with

$$\varrho p''(\varrho) + 2p'(\varrho) > 0\,. \tag{4.11}$$

This structure of the "pressure" p results from the fact that knowledge of this function is only of interest close to the vacuum, and the fact that $\varrho p(\varrho)$ is strictly convex.

Equation (4.10) is strictly hyperbolic for $\varrho > 0$ because it can be written as

$$\partial_t \boldsymbol{U} + \boldsymbol{A}(\boldsymbol{U})\partial_x \boldsymbol{U} = 0 \tag{4.12}$$

with $\boldsymbol{U} = (\varrho, v)^T$ and

$$\boldsymbol{A}(\boldsymbol{U}) = \begin{pmatrix} v & \varrho \\ 0 & v - \varrho p'(\varrho) \end{pmatrix}$$

with two real eigenvalues which are obviously different from each other as long as $\varrho > 0$.

The principles developed by Aw and Rascle after observing real traffic flows are as follows:

1. The system is expected to be strictly hyperbolic
2. The Riemann problem with arbitrary bounded nonnegative Riemann data U must lead to a solution where ϱ and v are nonnegative
3. If $U_\pm$ are the data for this Riemann problem, all waves connecting any state U to its left must lead to a shock speed $\lambda \le v$

69) A. Aw, M. Rascle: SIAM J. Appl. Math. **60** (2000) 916–983

4. Braking leads to shock waves, while acceleration leads to rarefaction waves that fulfill principle 3

5. There must be no continuous dependence of the solution to the Riemann problem on the initial data at $\varrho = 0$.

5
Real Fluids

5.1
The Navier–Stokes Equations Model

An ideal fluid was defined as being a material without friction between neighboring particles with different flow velocities, and thus without tangential forces along the surfaces of arbitrary volumes $W(t)$ picked out of the fluid flow. The mathematical model of ideal fluids can therefore only describe certain situations, e.g., low-speed flows, those with low densities, low viscosities or high temperatures, etc., and the results obtained from it can seriously contradict reality if these properties are not ensured. The force acting on a bridge pier parallel to the flow of a calmly flowing stream was mentioned when the Kutta–Zhukovsky formulas were presented in Section 1.3.

Thus, if the tangential forces generated by the viscosity play an important role in a real flow situation, they cannot be neglected in a mathematical model that is expected to describe the flow in a satisfactory manner.

Let us therefore try to revise the Euler equations model of ideal fluids by taking the viscosity forces into account too so that we can generate a model of real fluids.

Recall that in our scenario, viscosity always refers to friction between fluid particles,[70] and so these additional forces can only occur in the case of motion (i.e., for a nonvanishing flow velocity), in which case they lead to an additional term within the law of the conservation of momentum.

The right hand side of (1.5) consisted of the forces per unit volume $\boldsymbol{k}$ only, and so the forces for the volume $W(t)$ are therefore given by

$$\int_{W(t)} \boldsymbol{k} \, \mathrm{d}(x, y, z) \, .$$

The frictional forces generated by the fluid particles that are outside this volume but moving along its surface $\partial W(t)$ are obviously proportional to the area of $\partial W(t)$ and are directed tangentially, whereas the forces generated by the pressure of the external fluid are proportional to this area, too, but are directed perpendicular to $\partial W(t)$.

70) See (1.54) and its explanation

Mathematical Models of Fluid Dynamics. R. Ansorge and T. Sonar

ISBN: 978-3-527-40774-3

If $s = (s_1, s_2, s_3)^T$ are the forces per unit area that are generated by the outer fluid and act on the surface $\partial W(t)$ of an inner volume $W(t)$ arbitrarily picked out of the fluid, the relation

$$s = s(t, x, y, z, n) \;, \qquad (x, y, z) \in \partial W(t) \tag{5.1}$$

holds, where n is the outwardly directed unit vector normal on $\partial W(t)$ at $(x, y, z) \in \partial W(t)$.

Here, s depends linearly on n:

$$s = (\sigma_{ij})n \tag{5.2}$$

with σ_{ij}: $(x, y, z, t) \longrightarrow \sigma_{ij}(x, y, z, t)$.

Equation (5.2) follows from the fact that, within the law of the conservation of momentum, now written as

$$\int\limits_{W(t)} \left\{ \partial_t q + \left\langle \frac{1}{\varrho} q, \nabla \right\rangle q + \operatorname{div} \left(\frac{1}{\varrho} q \right) q \right\} \, d(x, y, z)$$
$$= \int\limits_{W(t)} k \, d(x, y, z) + \int\limits_{\partial W(t)} s \, do \;,$$

the volume terms vanish to a higher order than the surface terms when $W(t) \longrightarrow 0$.

Here, the *stress tensor* (σ_{ij}) is treated like a real (3, 3)-matrix.

Moreover, *Euler's law of vanishing momentum*, i.e.,

$$\int\limits_{W(t)} [r, k] \, d(x, y, z) + \int\limits_{\partial W(t)} [r, s] \, do = 0 \;, \tag{5.3}$$

$r = (x, y, z)^T$, yields the symmetry of the stress tensor:

$$\sigma_{ij} = \sigma_{ji} \;. \tag{5.4}$$

In the case of ideal fluids, vanishing friction must be modeled by $\sigma_{ij} = 0$ for $i \neq j$. Taking (5.2) into account, this leads to

$$\sigma_{11} = \sigma_{22} = \sigma_{33} =: -p \qquad \text{(pressure)}$$

because the vectors s and n are parallel in this situation. This is a slightly more general version of (1.5).

When real fluids are of interest, the terms $\sigma_{ij} (i \neq j)$ can be expected to be small, and the terms σ_{ii} will certainly differ from $-p$ by only a small amount. We therefore put

$$\tilde{\sigma}_{ij} = \begin{cases} \sigma_{ii} + p & \text{for} \quad i = j \\ \sigma_{ij} & \text{for} \quad i \neq j \,, \end{cases} \tag{5.5}$$

so that all of the $\tilde{\sigma}_{ij}$ are small.

Because of the symmetry of (σ_{ij}), the tensor $(\tilde{\sigma}_{ij})$ is also symmetric.

If the exterior forces $\boldsymbol{k}$ per unit volume are replaced by the forces $\hat{\boldsymbol{k}}$ per unit mass,

$$\int\limits_{W(t)} \left\{ \partial_t \boldsymbol{q} + \left\langle \frac{1}{\varrho}\boldsymbol{q}, \boldsymbol{\nabla} \right\rangle \boldsymbol{q} + \operatorname{div}\left(\frac{1}{\varrho}\boldsymbol{q}\right) \boldsymbol{q} - \varrho\hat{\boldsymbol{k}} \right\} \mathrm{d}(x, y, z)$$
$$= - \int\limits_{\partial W(t)} p\boldsymbol{n} \,\mathrm{do} + \int\limits_{\partial W(t)} (\tilde{\sigma}_{ik})\, \boldsymbol{n} \,\mathrm{do}$$

follows.

Obviously, if a fluid does not move or if neighboring particles do not move with different velocities, friction cannot occur. In modeling the flow, we therefore assume that the terms $\tilde{\sigma}_{ij}(i, j = 1, 2, 3)$ depend only on the components of the vectors $\boldsymbol{\nabla} u_i (i = 1, 2, 3)$. Here, we restrict ourselves to so-called *Newtonian fluids*, i.e., the terms $\tilde{\sigma}_{ij}$ depend on the components $\boldsymbol{\nabla} u_i$ homogeneously and linearly:

$$\tilde{\sigma}_{ij} = \sum_{\mu,\nu=1}^{3} \alpha_{\mu\nu}^{(i,j)} \partial_{x_\mu} u_\nu \,. \tag{5.6}$$

Thus, every $\tilde{\sigma}_{ij}$ is a function of the elements of the Jacobian $\boldsymbol{J}\boldsymbol{u}$:

$$\tilde{\sigma}_{ij} : \boldsymbol{J}\boldsymbol{u} \longrightarrow \tilde{\sigma}_{ij}(\boldsymbol{J}\boldsymbol{u}) \,,$$

and the $\alpha_{\mu\nu}^{(i,j)}$ are constants.

Analogously to the case of ideal fluids, where every axis is a main stress axis, we also expect relation (5.6) to be independent of orthogonal linear transformations in our model, i.e.,

$$\boldsymbol{A}^{\mathrm{T}} (\tilde{\sigma}_{ij})\, \boldsymbol{A} = \left(\tilde{\sigma}_{ij}\left(\boldsymbol{A}^{\mathrm{T}}\boldsymbol{J}\boldsymbol{u}\boldsymbol{A}\right)\right), \qquad \forall\, \boldsymbol{A} \in \mathbb{R}^{(3,3)} \quad \text{with} \quad \boldsymbol{A}^{\mathrm{T}}\boldsymbol{A} = \boldsymbol{I} \,. \tag{5.7}$$

It is trivial to show that the matrix $\boldsymbol{B} := \operatorname{div} \boldsymbol{u} \cdot \boldsymbol{I}$ is symmetric; it is also a diagonal matrix whose diagonal elements equal each other, so it shows the property

$$\boldsymbol{A}^{\mathrm{T}}\boldsymbol{B}\boldsymbol{A} = \boldsymbol{B} \,.$$

The elements b_{ij} of $\boldsymbol{B}$ depend on the elements of the Jacobian, i.e., $\boldsymbol{B} = (b_{ij}(\boldsymbol{J}\boldsymbol{u}))$. Because $\operatorname{div} \boldsymbol{u} = \operatorname{trace} \boldsymbol{J}\boldsymbol{u}$ and because traces of matrices are invariant to orthogonal transformations,

$$\left(b_{ij}\left(\boldsymbol{A}^{\mathrm{T}}\boldsymbol{J}\boldsymbol{u}\boldsymbol{A}\right)\right) = \left(b_{ij}(\boldsymbol{J}\boldsymbol{u})\right) = \boldsymbol{B}$$

also follows.

Thus, $\boldsymbol{B}$ fulfills all of the requirements for $(\tilde{\sigma}_{ij})$. However,

$$\boldsymbol{D} := \boldsymbol{J}\boldsymbol{u} + (\boldsymbol{J}\boldsymbol{u})^{\mathrm{T}} =: \boldsymbol{D}\,(\boldsymbol{J}\boldsymbol{u})$$

is also symmetric, it depends on the elements of the Jacobian in a homogeneous and linear way, and it shows the property

$$\begin{aligned} \boldsymbol{A}^{\mathrm{T}}\boldsymbol{D}\boldsymbol{A} &= \boldsymbol{A}^{\mathrm{T}}\boldsymbol{J}\boldsymbol{u}\boldsymbol{A} + \boldsymbol{A}^{\mathrm{T}}\,(\boldsymbol{J}\boldsymbol{u})^{\mathrm{T}}\,\boldsymbol{A} \\ &= \boldsymbol{A}^{\mathrm{T}}\boldsymbol{J}\boldsymbol{u}\boldsymbol{A} + \left(\boldsymbol{A}^{\mathrm{T}}\boldsymbol{J}\boldsymbol{u}\boldsymbol{A}\right)^{\mathrm{T}} \,, \\ \text{i.e.,} \qquad \boldsymbol{A}^{\mathrm{T}}\boldsymbol{D}(\boldsymbol{J}\boldsymbol{u})\boldsymbol{A} &= \boldsymbol{D}\left(\left(\boldsymbol{A}^{\mathrm{T}}\boldsymbol{J}\boldsymbol{u}\boldsymbol{A}\right)\right) . \end{aligned}$$

Thus, $\boldsymbol{D}$ also fulfills all of the demands on $(\tilde{\sigma}_{ij})$, as does $\lambda\boldsymbol{B} + \eta\boldsymbol{D}$ with the arbitrary constants λ, η. It can be shown that all of these types of matrices satisfy our demands. Hence,

$$\int\limits_{W(t)} \left\{ \partial_t \boldsymbol{q} + \left\langle \frac{1}{\varrho}\boldsymbol{q}, \nabla \right\rangle \boldsymbol{q} + \operatorname{div}\left(\frac{1}{\varrho}\boldsymbol{q}\right)\boldsymbol{q} - \varrho\hat{\boldsymbol{k}} \right\} \mathrm{d}(x, y, z) - \int\limits_{\partial W(t)} (-p\boldsymbol{I} + \lambda\boldsymbol{B} + \eta\boldsymbol{D})\,\boldsymbol{n}\,\mathrm{d}o = 0\ . \tag{5.8}$$

By means of the divergence theorem,

$$\begin{aligned} \int\limits_{\partial W(t)} p\boldsymbol{n}\,\mathrm{d}o &= \int\limits_{W(t)} \nabla p\,\mathrm{d}(x, y, z) \\ \text{and} \int\limits_{\partial W(t)} \boldsymbol{B}\boldsymbol{n}\,\mathrm{d}o &= \int\limits_{W(t)} \begin{pmatrix} \operatorname{div} \boldsymbol{b}_1 \\ \operatorname{div} \boldsymbol{b}_2 \\ \operatorname{div} \boldsymbol{b}_3 \end{pmatrix} \mathrm{d}(x, y, z) \end{aligned}$$

follow, where the $\boldsymbol{b}_i$ are the row vectors of $\boldsymbol{B}$, i.e.,

$$\int\limits_{\partial W(t)} \boldsymbol{B}\boldsymbol{n}\,\mathrm{d}o = \int\limits_{W(t)} \nabla(\operatorname{div}\boldsymbol{u})\,\mathrm{d}(x, y, z)\ ,$$

and similarly

$$\int\limits_{\partial W(t)} \boldsymbol{D}\boldsymbol{n}\,\mathrm{d}o = \int\limits_{W(t)} \begin{pmatrix} \operatorname{div} \boldsymbol{d}_1 \\ \operatorname{div} \boldsymbol{d}_2 \\ \operatorname{div} \boldsymbol{d}_3 \end{pmatrix} \mathrm{d}(x, y, z)$$

where

$$\begin{aligned} \operatorname{div}\boldsymbol{d}_1 &= 2\partial_x(\partial_x u_1) + \partial_y(\partial_y u_1 + \partial_x u_2) + \partial_z(\partial_z u_1 + \partial_x u_3) \\ &= \Delta u_1 + \partial_{xx} u_1 + \partial_{xy} u_2 + \partial_{xz} u_3 \\ &= \Delta u_1 + \partial_x(\operatorname{div}\boldsymbol{u})\ . \end{aligned}$$

Analogous formulae hold for $\operatorname{div}\boldsymbol{d}_2$ and $\operatorname{div}\boldsymbol{d}_3$.

Equation (5.8) therefore becomes

$$\int\limits_{W(t)} \left\{ \partial_t \boldsymbol{q} + \left\langle \frac{1}{\varrho}\boldsymbol{q}, \nabla \right\rangle \boldsymbol{q} + \operatorname{div}\left(\frac{1}{\varrho}\boldsymbol{q}\right)\boldsymbol{q} - \varrho\hat{\boldsymbol{k}} + \nabla p - \lambda\nabla(\operatorname{div}\boldsymbol{u}) - \eta\Delta\boldsymbol{u} - \eta\nabla(\operatorname{div}\boldsymbol{u}) \right\} \mathrm{d}(x, y, z) = 0$$

for each part $W(t)$ picked arbitrarily out of the volume of the fluid, i.e.,

$$\partial_t(\varrho\boldsymbol{u}) + \langle \boldsymbol{u}, \nabla\rangle(\varrho\boldsymbol{u}) + (\varrho\operatorname{div}\boldsymbol{u})\boldsymbol{u} - \varrho\hat{\boldsymbol{k}} + \nabla p - (\lambda + \eta)\nabla(\operatorname{div}\boldsymbol{u}) - \eta\Delta\boldsymbol{u} = 0$$

or

$$\partial_t \varrho \boldsymbol{u} + \varrho \partial_t \boldsymbol{u} + \langle \boldsymbol{u}, \nabla \varrho \rangle \boldsymbol{u} + \varrho \langle \boldsymbol{u}, \nabla \rangle \boldsymbol{u} + (\varrho \operatorname{div} \boldsymbol{u}) \boldsymbol{u} - \varrho \hat{\boldsymbol{k}} + \nabla p$$
$$- (\lambda + \eta) \nabla (\operatorname{div} \boldsymbol{u}) - \eta \Delta \boldsymbol{u} = 0 \, .$$

Because of $\varrho \operatorname{div} \boldsymbol{u} + \langle \boldsymbol{u}, \nabla \varrho \rangle = \operatorname{div}(\varrho \boldsymbol{u})$, the relation

$$\{\partial_t \varrho + \operatorname{div}(\varrho \boldsymbol{u})\} \, \boldsymbol{u} + \varrho \partial_t \boldsymbol{u} + \varrho \langle \boldsymbol{u}, \nabla \rangle \boldsymbol{u} - \varrho \hat{\boldsymbol{k}} + \nabla \varrho - (\lambda + \eta) \nabla (\operatorname{div} \boldsymbol{u}) - \eta \, \Delta \boldsymbol{u} = 0$$

follows. Taking the continuity equation into account, and dividing by ϱ, we then obtain the so-called *Navier*[71]*–Stokes*[72] *equations:*

$$\boxed{\partial_t \boldsymbol{u} + \langle \boldsymbol{u}, \nabla \rangle \boldsymbol{u} - \frac{\lambda + \eta}{\varrho} \nabla (\operatorname{div} \boldsymbol{u}) - \frac{\eta}{\varrho} \Delta \boldsymbol{u} = -\frac{1}{\varrho} \boldsymbol{\nabla} \, p + \hat{\boldsymbol{k}}} \tag{5.9}$$

Here, η is called the *first viscosity coefficient.* Of course, historically speaking, there were several arguments that led to the Navier–Stokes equations, which is why λ is not called the *second viscosity coefficient,* but rather $\xi := \lambda + \frac{2}{3}\eta$. These coefficients are constants that depend on the fluid used as well as its temperature. The unit of measurement is $\text{g} \cdot \text{cm}^{-1} \cdot \text{s}^{-1}$ = Poise (cf. footnote 74).

When considering incompressible flows characterized by $\operatorname{div} \boldsymbol{u} = 0$, λ does not occur in the equations, and if the exterior forces $\hat{\boldsymbol{k}}$ do not play a role either, the Navier–Stokes equations are reduced to

$$\partial_t \boldsymbol{u} + \langle \boldsymbol{u}, \nabla \rangle \boldsymbol{u} - \nu \Delta \boldsymbol{u} + \frac{1}{\varrho} \nabla p = 0 \, , \tag{5.10}$$

where $\nu = \dfrac{\eta}{\varrho}$ is called the *kinematic viscosity.*

By similar arguments, the law describing the conservation of energy takes the following form for viscous flows:

$$\partial_t E + \operatorname{div} \left((E + p) \, \boldsymbol{u} \right) - \lambda \cdot \operatorname{div}(\operatorname{div} \boldsymbol{u} \cdot \boldsymbol{u}) - \eta \cdot \operatorname{div} \left(\left(J\boldsymbol{u} + (J\boldsymbol{u})^{\mathrm{T}} \right) \boldsymbol{u} \right) = 0 \, .$$

Some of the terms within this equation can be simplified, e.g.,

$$\operatorname{div}(\operatorname{div} \boldsymbol{u} \cdot \boldsymbol{u}) = \langle \nabla (\operatorname{div} \boldsymbol{u}), \boldsymbol{u} \rangle + (\operatorname{div} \boldsymbol{u})^2$$

etc.

The Navier–Stokes equations model has been shown by experiments to yield excellent results.

A feel for the magnitude of kinematic viscosity and its sensitivity to temperature is given by Table 5.1.

71) Claude Louis Marie Henri Navier (1785–1836); Dijon, Paris

72) George Gabriel Stokes (1819–1903); Cambridge

Table 5.1 Values of the kinematic viscosity.

Fluid	Temperature [°C]	Kinematic viscosity $[\mathrm{cm}^2/\mathrm{s}]$
Mercury	0	0.00125
	10	0.00123
	20	0.00117
Air	0	0.133
	10	0.140
	20	0.143
Water	0	0.0178
	10	0.0130
	20	0.0101
Machine oil	0	7.34
(brand-dependent; approx. values shown)	20	3.82

In order to experimentally study the behavior of a real ship in the ocean or a wing passing through the air, etc., engineers normally use a small model of the ship that is dipped into a wave canal or a small model of the wing that is placed in a wind tunnel, respectively. Let us attempt to find the conditions under which the results of the experiments agree with reality. To answer this question, it is advisable to use dimensionless quantities.

For convenience, let us consider the case of an incompressible flow.

Assume that $\boldsymbol{u}_0 = (u_0, 0, 0)^T$ is the flow velocity measured far away from the solid, i.e., from the hull of the ship, from the wing etc. Let L be a characteristic length of the solid, e.g., the length of the ship, the length of the wing's airfoil, etc., let T be the time the fluid needs to transit the length L at velocity $\boldsymbol{u}_0$, and let $P = \frac{\varrho}{2} u_0^2$ be the dynamic pressure in a great distance of the solid.

Let us consider the particular case where the exterior forces per unit volume can be neglected. We now introduce the dimensionless quantities

$$\tilde{\boldsymbol{u}} = \frac{1}{u_0}\boldsymbol{u}\,, \quad \tilde{t} = \frac{t}{T}\,, \quad \tilde{p} = \frac{p}{2P}\,, \quad \tilde{x} = \frac{x}{L}\,, \quad \tilde{y} = \frac{y}{L}\,, \quad \tilde{z} = \frac{z}{L}\,.$$

The incompressible Navier–Stokes equations then become

$$\partial_{\tilde{t}}\tilde{\boldsymbol{u}} + \langle \tilde{\boldsymbol{u}}, \tilde{\nabla} \rangle \tilde{\boldsymbol{u}} + \tilde{\nabla}\tilde{p} - \frac{1}{\mathrm{Re}}\tilde{\Delta}\tilde{\boldsymbol{u}} = 0\,, \tag{5.11}$$

where

$$\mathrm{Re} := \frac{\varrho u_0 L}{\eta} = \frac{u_0 L}{\nu} \tag{5.12}$$

is the so-called *Reynolds number* and

$$\tilde{\nabla} := \left(\partial_{\tilde{x}}, \partial_{\tilde{y}}, \partial_{\tilde{z}}\right)^{\mathrm{T}}\,, \quad \tilde{\Delta} = \partial_{\tilde{x}\tilde{x}} + \partial_{\tilde{y}\tilde{y}} + \partial_{\tilde{z}\tilde{z}}\,.$$

Thus, the experiment coincides with the real-world behavior of the solid when placed in the flow if the Reynolds numbers of both situations coincide. This leaves a great deal of freedom in the particular choices of $\boldsymbol{u}_0$, L, ν that can be used in the experiment. The flows are then said to be *similar*.

In the case of $\eta \neq 0$, the Laplace operator occurs in (5.10), so this system of equations now becomes a parabolic system if the flow is instationary or an elliptic system if a higher dimensional stationary flow is present, respectively.

The continuity equation as well as the energy conservation law that is already known from the model of ideal fluid flows remain unaffected by the viscosity and are therefore still valid. This also holds for the additional equation of state when gas flows are being investigated.

There are only a few situations where the Euler equations or the Navier–Stokes equations can be explicitly solved by well-known functions. In all other situations, they must be solved approximately using numerical procedures.

Two of the problems which can be treated explicitly will be studied in the next section, and both problems are examples of stationary flows.

Remark

The Navier–Stokes equations do not consider the additional buoyancy phenomenona that arise in compressible gas flows from density variations caused by temperature differences between different parts of the gas. In order to account for these phenomenona, the Navier–Stokes equations have to be enriched with additional terms. If the enriched equations are also written in a dimensionless form, further dimensionless parameters besides the Reynolds number occur, in particular the Prandtl number

$$\mathrm{Pr} := \frac{\nu}{\kappa} \varrho c_p$$

where κ is the specific thermal conductivity. An example of such a scenario is the flow of hot gases through pipes, e.g., tunnel fires, as studied in Section 5.7.

5.2 Drag Force and the Hagen–Poiseuille Law

Viscous flows can be *laminar* or *turbulent*. The flow is laminar if neighboring fluid particles move more or less parallel to each other along stable trajectories, whereas turbulent flows are characterized by a disordered flow superimposed on the main flow. Thus, the turbulent flow is an instability phenomenon. When Reynolds studied the flow of liquids through small glass tubes, he noticed that the transition from the laminar to the turbulent state of the flow takes place in a very abrupt way, and depends on the parameter that is now called the Reynolds number (cf. (5.11)).

The transition to instability will be briefly discussed in the fifth section of this chapter. In this section, the flows are assumed to be laminar.

Our first simple application of the Navier–Stokes equations concerns a laminar viscous stationary flow in the y-direction with a small velocity parallel to a given area δF of a plain solid plate, where this area is part of the (y, z)-plane. Let the fluid be incompressible in the sense of ϱ = const. Because of the friction between the moving fluid particles and the fluid particles that are stuck to the rigid body, in accordance with the no-slip condition, there is a force acting on δF. Obviously, this force has only one component in the y-direction, and the velocity is independent of the space variables y and z:

$$\boldsymbol{u} = (0,\ u(x),\ 0)^T \quad \text{with} \quad u(0) = 0 \quad \text{(no-slip condition)}\ .$$

The undisturbed velocity of the fluid that passes over the surface area at a great distance is assumed to be constant, i.e., $\partial_x u(x) \to 0$ for $x \to \infty$.

Compared with (5.10), the forces per mass unit occuring within the Euler equations are now enriched by the additional term $\nu \Delta \boldsymbol{u}$, so that the forces per volume unit are now modeled by

$$\boldsymbol{w} = \begin{pmatrix} 0 \\ \eta \partial_{xx} u(x) \\ 0 \end{pmatrix} . \tag{5.13}$$

The resistance of the relevant area (which acts in the direction opposite to the flow and arises from friction) or vice versa the *drag force* that acts on the plate due to the friction of the fluid is therefore

$$\delta F \int_0^\infty \boldsymbol{w}\, \mathrm{d}x = \begin{pmatrix} 0 \\ -\eta u_x(0) \delta F \\ 0 \end{pmatrix} .$$

Hence, the drag related to the friction of a laminar flow is given by

$$W = \eta \partial_x u(0) \delta F\ . \tag{5.14}$$

Coming back to Reynolds' experiments, as an example let us now study the one-dimensional laminar stationary cylindrically symmetric flow of an incompressible liquid through a cylindrical pipe of radius R where gravitational force is neglected, where other outer forces do not occur, and where the no-slip condition $\boldsymbol{u} = 0$ holds at the interior surface of the pipe (cf. Fig. 5.2).

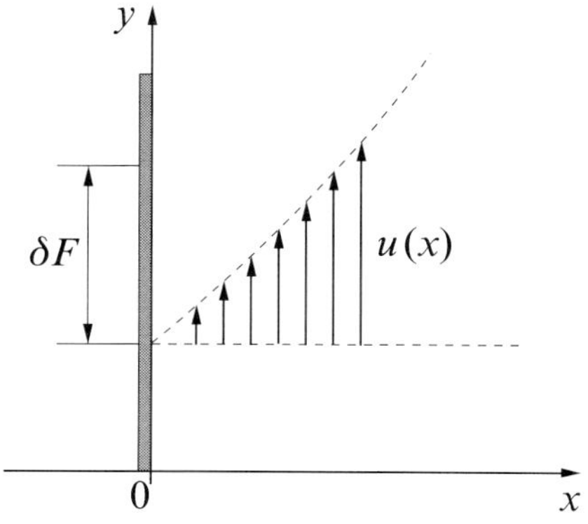

Fig. 5.1 Flow along a planar membrane.

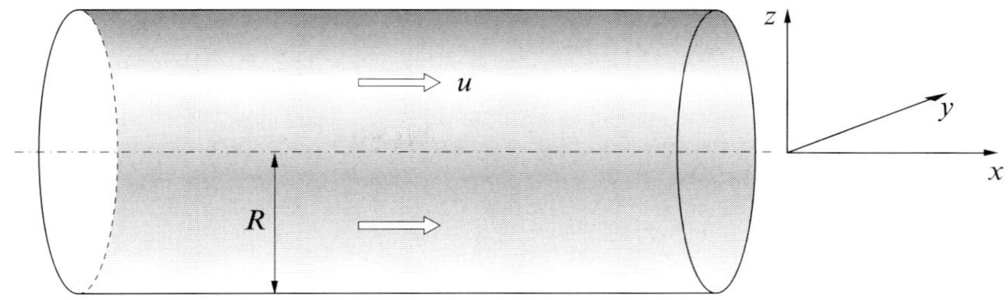

Fig. 5.2 Laminar pipe flow.

Because of $\boldsymbol{u} = (u, 0, 0)^T$, the relations

$$\langle \boldsymbol{u}, \nabla \rangle \, \boldsymbol{u} = (u\partial_x u, 0, 0)^T \,,$$

$$\Delta \boldsymbol{u} = (\Delta u, 0, 0)^T \quad \text{and} \quad \operatorname{div} \boldsymbol{u} = \partial_x u$$

follow.

Taking the continuity equation $\operatorname{div} \boldsymbol{u} = 0$ into account, we find that $\partial_x u \equiv 0$, and so $\partial_{xx} u \equiv 0$ too. Hence, (5.8) reduces to

$$-\frac{\eta}{\varrho}\begin{pmatrix} \Delta u \\ 0 \\ 0 \end{pmatrix} + \frac{1}{\varrho}\begin{pmatrix} \partial_x p \\ \partial_y p \\ \partial_z p \end{pmatrix} = \begin{pmatrix} 0 \\ 0 \\ 0 \end{pmatrix} . \tag{5.15}$$

This leads to

$$\partial_y p \equiv \partial_z p \equiv 0 \,, \quad \text{i.e.,} \quad p = p(x) \,,$$

and therefore to

$$\Delta u = \partial_{yy} u + \partial_{zz} u = \frac{\partial_x p}{\eta} \,. \tag{5.16}$$

The left side of (5.16) is represented by a function that depends only on the variables y and z, whereas the right side depends only on x. Therefore, $\partial_x p$ must be constant.

The particular type of flow considered here can therefore only occur if p depends linearly on x.

If polar coordinates (r, φ) are introduced in the (y, z)-plane, and if the cylindrical symmetry is respected, i.e., $u = u(r)$ independent of φ, the well-known transformation rule

$$\Delta u = \frac{1}{r}\partial_r \left(r\partial_r u\right) + \frac{1}{r^2}\partial_{\varphi\varphi} u$$

transforms (5.16) into

$$\partial_r \left(r\partial_r u\right) = \frac{\partial_x p}{\eta} r \,.$$

Because p is independent of (y, z), and hence independent of (r, φ), integration leads to

$$r\partial_r u = \frac{\partial_x p}{\eta}\frac{r^2}{2} + c_1 \tag{5.17}$$

with an integration constant c_1.

Besides the no-slip condition $u(R) = 0$, the boundary condition

$$|(\partial_r u)_{r=0}| < \infty$$

must also be fulfilled, so c_1 has to vanish.

If (5.17) is now integrated,

$$u(r) = \frac{\partial_x p}{2\eta}\frac{r^2}{2} + c_2$$

results, where the constant c_2 must be chosen in such a way that the no-slip condition is fulfilled. This finally yields the so-called *Hagen–Poiseuille law*

$$u(r) = -\frac{\partial_x p}{4\eta}\left(R^2 - r^2\right). \tag{5.18}$$

This law exhibits a parabolic velocity distribution for every fixed x, and the flow follows the direction of decreasing pressure (cf. Fig. 5.3).

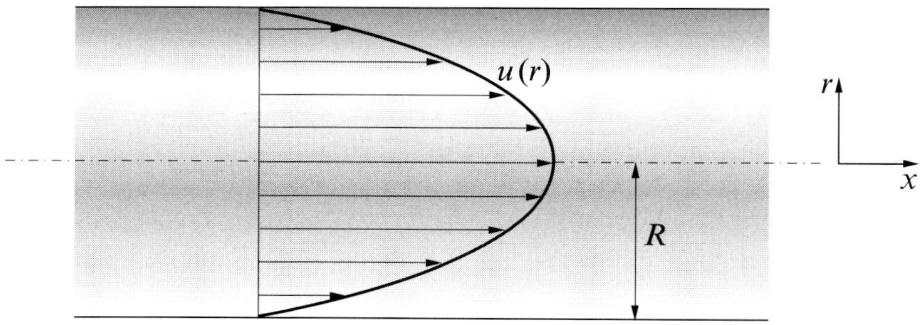

Fig. 5.3 The Hagen–Poiseuille flow through a pipe.

We can now immediately determine the mass that flows through the cross-section Q of the tube per second:

$$M = \int_Q \varrho u \, \mathrm{d}Q = -\int_0^{2\pi}\int_0^R \frac{\varrho\partial_x p}{4\eta}\left(R^2 - r^2\right) r \, \mathrm{d}r \, \mathrm{d}\varphi = -\frac{\pi\varrho\partial_x p}{8\eta}R^4 , \tag{5.19}$$

which is proportional to the fourth power of the radius R, though the cross-section is only proportional to the second power of the radius.

Remark

This mass flow was experimentally discovered by the Prussian civil service hydraulic engineer Hagen[73] in 1839 without any knowledge of (5.18), and also (independently) by the French physiologist Poiseuille[74] in 1841, when he was interested in blood flow. It was also Poiseuille who discovered Eq. (5.19). The parabolic velocity distribution was first mentioned in 1845 by Stokes.

The experiments of Hagen and of Poiseuille were performed with small tube diameters (0.015–3.00 mm). They confirmed Eq. (5.19), and so the law (5.18) is now named after them. Reynolds used glass tubes with greater diameters (up to 10 mm); he also added some dyes to the liquids he worked with and noted that there is a critical value of the parameter which is now called the Reynolds number, and this *critical Reynolds number* is characteristic of the transition from laminar flow to turbulent flow. He found that $\mathrm{Re}_{\mathrm{cr}} = 1160$, and recognized that the Hagen–Poiseuille law loses its validity in regions of turbulence: the mean flow velocity decreases due to a higher flow resistance.

The drag force of the laminar Hagen–Poiseuille flow in a section of a cylindrical pipe of length L can be computed via (5.14) and of (5.18) in the following way:

$$W = \eta 2\pi RL\,(\partial_r u)_{r=R} = \eta 2\pi RL\frac{\partial_x pR}{2\eta} = \pi R^2 \Delta p\,, \tag{5.20}$$

where Δp denotes the pressure difference between the two ends of the particular part of the pipe under consideration.

There are some other cross-sections of ducts for which the Navier–Stokes equations can be solved analytically under the same flow assumptions as used in the Hagen–Poiseuille case (steady, laminar, incompressible, viscous), such as annuli, equilateral triangles, etc.[75]

Situations similar to the flow of a liquid through narrow ducts occur for other flows through narrow gaps or slits, e.g., in investigations of the flow of lubricating oils in lubrication research – termed *tribology*.

As already mentioned, there are only a few examples where the Navier–Stokes equations can be solved exactly. This led to efforts to simplify these equations wherever possible. Some of the most important developments concerning such simplifications for small and large Reynolds numbers will be briefly discussed in the next sections. Also, for large Reynolds numbers, we do not go back to the Euler equations but instead preserve some of the structures of the Navier–Stokes equations and study other suitable simplifications, and the no-slip condition will also be retained for large Reynolds numbers. The small Reynolds number case was particularly thoroughly investigated by Stokes and led to a model equation that is now called the *Stokes approximation* (cf. (5.25)). On the other hand, the theory and practical applications of fluid dynamics for large Reynolds numbers was significantly advanced by Prandtl[76] in his *boundary layer theory* and his *wing theory*.

73) Gotthilf Heinrich Ludwig Hagen (1797–1884); Berlin

74) Jean Louis Marie Poiseuille (1799–1869); Paris

75) C.Y. Wang: Ann. Rev. Fluid Mech. **23** (1991) 159–177

76) Ludwig Prandtl (1875–1953); Hannover, Göttingen

5.3 Stokes Approximation and Artificial Time

The difficulties involved in solving stationary or instationary Navier–Stokes problems or Euler problems are mainly due to the nonlinearity of the convection term $\langle \boldsymbol{u}, \nabla \rangle \boldsymbol{u}$.

Another difficulty, which is particularly associated with the case of an incompressible instationary flow, is the fact that the pressure p that must be determined along with the velocity $\boldsymbol{u}$ is not represented in the continuity equation by its time derivative, e.g., by an additional term $\kappa \partial_t p$ ($\kappa = \text{const}$). If it would have been represented by its time derivative, and if a suitable initial pressure p_0 would have been prescribed, the instationary problem could be reformulated as a genuine initial boundary value problem with the differential equation

$$\partial_t \varphi - L[\varphi] = 0 \quad \text{with} \quad \varphi = (u_1, u_2, u_3, p)^T \tag{5.21}$$

which could be treated theoretically as well as numerically using appropriate techniques.

Let us assume that the solutions of instationary Navier–Stokes problems converge for $t \longrightarrow \infty$ in each case towards the solution of the corresponding stationary problem, independent of the initial function $\boldsymbol{u}_0$.

If this also holds for the instationary problem (5.21) consisting of the Navier–Stokes equations and of a continuity equation enriched with a physically unjustified additional time derivative of p, the solution of this manipulated problem must coincide with the stationary solution of the original Navier–Stokes problem for $t \longrightarrow \infty$ because of

$$p_t \longrightarrow 0 \quad \text{for} \quad t \longrightarrow \infty \, .$$

This leads to the following idea for a method of solving the stationary Navier–Stokes problem. Introduce an *artificial time* τ and enrich the stationary Navier–Stokes equations as well as the continuity equation by adding linear combinations of certain terms in such a way that, after the addition of a more or less arbitrary initial function, a complete instationary initial or initial boundary value problem is created with a solution that is expected to be able to nullify the additional artificial terms for $\tau \longrightarrow \infty$. In engineering, this concept is called the *method of artificial compressibility* and is occasionally realized numerically.

Such an artificially enriched stationary problem might appear as follows:

$$\left.\begin{aligned} -\partial_\tau \boldsymbol{u} &= \langle \boldsymbol{u}, \nabla \rangle \boldsymbol{u} + \nabla p - \frac{1}{\text{Re}} \Delta \boldsymbol{u} + \alpha \nabla \operatorname{div} \boldsymbol{u} \\ -\partial_\tau p &= \beta^2 \operatorname{div} \boldsymbol{u} \end{aligned}\right\} \quad \text{in} \quad \Omega \, , \tag{5.22}$$

$$\boldsymbol{u} = \boldsymbol{u}_0 \quad \text{on} \quad \partial\Omega \, ,$$

$$\boldsymbol{u} = \boldsymbol{u}^{[0]} \, , \quad p = p_0 \quad \text{for} \quad \tau = 0 \, ,$$

where $\boldsymbol{u}^{[0]}$ is assumed to coincide with $\boldsymbol{u}_0$ on $\partial\Omega$. $\alpha > 0$ and $\beta \neq 0$ are arbitrary constants. They should be chosen in such a way that the numerical treatment of the artificially enriched problem produces a method that is as effective as possible.

Of course, the expected property of the artificially added terms – that they will vanish as τ increases – should be guaranteed by a mathematical proof.

From a numerical point of view, this strategy can lead to an iterative procedure for approximately solving the stationary boundary value problem in the following way.

Use the more or less arbitrarily added artificial initial function as an initial approximation, i.e.,

$$\varphi^{[0]} := \begin{pmatrix} \boldsymbol{u}^{[0]} \\ p_0 \end{pmatrix} . \tag{5.23}$$

Then discretize (5.22) with respect to the artificial time variable τ, using a suitable step size $\Delta_n \tau$ in each step, e.g., an equidistant step size $\Delta\tau$.

An example of an actual semidiscrete approximate equation is the explicit Euler method

$$\varphi^{[n+1]} = \varphi^{[n]} - \Delta_n \tau \cdot L\left[\varphi^{[n]}\right] \quad (n = 0, 1, 2, \cdots) \tag{5.24}$$

formulated in the style of (5.21). Here, every $\boldsymbol{u}^{[n]}$ is expected to fulfill the boundary conditions given in (5.22). $\varphi^{[n]}$ will then not only be interpreted as an approximate solution of the artificial instationary problem at its nth artificial time step, but also as an approximate solution of the stationary problem after the nth iteration step. If the iteration converges sufficiently fast, it will be stopped after a certain number of steps n_0, and $\varphi^{[n_0]}(x, y, z)$ will then be considered the final approximate solution to the stationary problem.

Initially, this method was only used and justified heuristically by engineers, who asked the mathematicians for a mathematical justification too. This justification could then – at least partially – be provided by a proof of the fact that, for the Stokes approximation, the solution of (5.22) actually converges for $\tau \to \infty$ to the solution of the stationary problem.[77] Here, the truncation errors introduced by additional space discretizations are ignored.

The Stokes approximation is developed from the Navier–Stokes equations by omitting the convective term. This seems a reasonable omission for incompressible flows with small Reynolds numbers. Namely, if (5.11) is written as

$$\mathrm{Re} \cdot \partial_{\tilde{t}} \tilde{\boldsymbol{u}} + \mathrm{Re} \cdot \langle \tilde{\boldsymbol{u}}, \tilde{\nabla} \rangle \tilde{\boldsymbol{u}} + \mathrm{Re} \cdot \tilde{\nabla}\, \tilde{p} - \tilde{\Delta}\, \tilde{\boldsymbol{u}} = 0 \ ,$$

one can see that the linear diffusion term dominates over the convection term.

If we omit the tildes and restrict ourselves to the stationary problem, only

$$\mathrm{Re} \cdot \nabla\, p - \Delta \boldsymbol{u} = 0 \tag{5.25}$$

remains upon ignoring the convection term. Finally, we replace $\mathrm{Re} \cdot p$ with a new quantity p and enrich the system in the way presented in (5.22) for example. The

77) St. Fellehner: Dissertation, Hamburg 1995

system is then found to take the form

$$\left.\begin{aligned}\partial_\tau \boldsymbol{u} + \nabla p &= \Delta \boldsymbol{u} + \alpha \nabla \operatorname{div} \boldsymbol{u} \\ \partial_\tau p + \beta^2 \operatorname{div} \boldsymbol{u} &= 0\end{aligned}\right\} \quad \text{in} \quad \Omega , \tag{5.26}$$

$$\boldsymbol{u} = 0 \quad \text{on} \quad \partial\Omega ,$$

$$\boldsymbol{u} = \boldsymbol{u}^{[0]} , \quad p = p_0 \quad \text{for} \quad \tau = 0 ,$$

where the linear structure of (5.25) is also taken into account. Because of this structure, it is sufficient to show the convergence of $\boldsymbol{u}$ for increasing values of τ to the stationary solution in the case of vanishing exterior forces and homogeneous boundary conditions.

The velocity part of the stationary solution of (5.26) then obviously vanishes, i.e.,

$$\boldsymbol{u} \equiv 0 ,$$

and the following theorem holds:

Theorem 5.1

The solution $\boldsymbol{u}$ of (5.26)[78] converges in the sense of the L_2-norm for $\tau \to \infty$ to the zero function.[79]

Proof: Using the notation

$$\boldsymbol{J u} \cdot \boldsymbol{v} = (\langle \nabla u_1, \boldsymbol{v}\rangle, \langle \nabla u_2, \boldsymbol{v}\rangle, \cdots, \langle \nabla u_d, \boldsymbol{v}\rangle)^{\mathrm{T}} ,$$

$$\langle \boldsymbol{Ju}, \boldsymbol{Jv}\rangle = \sum_{i,j} \partial_{x_j} u_i \partial_{x_j} v_i ,$$

$$(p, q) = \int_\Omega pq \,\mathrm{d}\boldsymbol{x} , \qquad \|p\|_\Omega = \sqrt{(p, p)} \qquad \text{in} \quad L_2(\Omega) ,$$

$$(\boldsymbol{u}, \boldsymbol{v}) = \int_\Omega \langle \boldsymbol{u}, \boldsymbol{v}\rangle \,\mathrm{d}\boldsymbol{x} , \qquad \|\boldsymbol{u}\|_\Omega = \sqrt{(\boldsymbol{u}, \boldsymbol{u})} \qquad \text{in} \quad L_2^d(\Omega) ,$$

$$(\boldsymbol{Ju}, \boldsymbol{Jv}) = \int_\Omega \langle \boldsymbol{Ju}, \boldsymbol{Jv}\rangle \,\mathrm{d}\boldsymbol{x} , \quad \|\boldsymbol{Ju}\|_\Omega = \sqrt{(\boldsymbol{Ju}, \boldsymbol{Ju})} \quad \text{in} \quad L_2^{d\times d}(\Omega) ,$$

taking

$$\operatorname{div}(\boldsymbol{Ju} \cdot \boldsymbol{v}) = \langle \Delta \boldsymbol{u}, \boldsymbol{v}\rangle + \langle \boldsymbol{Ju}, \boldsymbol{Jv}\rangle ,$$

into account or – using Gauss' divergence theorem – the relation

$$(\Delta \boldsymbol{u}, \boldsymbol{v}) = \int_{\partial\Omega} \left\langle \frac{\partial \boldsymbol{u}}{\partial n}, \boldsymbol{v} \right\rangle \,\mathrm{d}o - (\boldsymbol{Ju}, \boldsymbol{Jv}) , \tag{5.27}$$

and finally also taking into account the fact that the integral along the bound-

78) For convenience, we omit considerations concerning the behavior of p.

79) Here, it is assumed that there is a unique smooth solution to the artificial instationary problem.

ary vanishes, (5.26) leads to

$$\begin{aligned}
\frac{1}{2}\frac{\mathrm{d}}{\mathrm{d}\tau}\left(\|\boldsymbol{u}\|_\Omega^2+\frac{1}{\beta^2}\|p\|_\Omega^2\right) &= (\boldsymbol{u},\boldsymbol{u}_\tau)+\frac{1}{\beta^2}(p,\partial_\tau p)\\
&= (\boldsymbol{u},-\nabla p)+(\boldsymbol{u},\Delta\boldsymbol{u})+(\boldsymbol{u},\alpha\nabla(\operatorname{div}\boldsymbol{u}))\\
&\quad+\frac{1}{\beta^2}\left(p,-\beta^2\operatorname{div}\boldsymbol{u}\right)\\
&= -\|\boldsymbol{J}\boldsymbol{u}\|_\Omega^2-\alpha\|\operatorname{div}\boldsymbol{u}\|_\Omega^2\,.
\end{aligned} \tag{5.28}$$

Thus,

$$\sigma(\tau) := \left(\|\boldsymbol{u}\|_\Omega^2+\frac{1}{\beta^2}\|p\|_\Omega^2\right)$$

decreases monotonously, and more strongly as α increases.[80] Hence, $\|\boldsymbol{u}\|_\Omega$ and $\|p\|_\Omega$ are bounded such that

$$\sigma_\infty := \lim_{\tau\to\infty}\sigma(\tau)$$

exists.

The Poincaré inequality[81] then yields in $L_2^d(\Omega)$

$$\|\boldsymbol{u}\|_\Omega^2 \le c_1\|\boldsymbol{J}\boldsymbol{u}\|_\Omega^2$$

where the constant $c_1 > 0$ depends only on the domain Ω. Equation (5.27) therefore leads to

$$\sigma'(\tau) = \frac{\mathrm{d}}{\mathrm{d}\tau}\left(\|\boldsymbol{u}\|_\Omega^2+\frac{1}{\beta^2}\|p\|_\Omega^2\right) \le -c\|\boldsymbol{u}\|_\Omega^2 \tag{5.29}$$

where $c = \dfrac{2}{c_1} > 0$.

Formula (5.29) implies

$$\sigma(\tau_0)-\sigma_\infty \ge c\int_{\tau_0}^{\infty}\|\boldsymbol{u}\|_\Omega^2\,\mathrm{d}\tau \tag{5.30}$$

so that, in particular, the integral on the right hand side exists, i.e., $\|\boldsymbol{u}\|_\Omega \in L_2(\mathbb{R}^+)$.

However,

$$\frac{\mathrm{d}}{\mathrm{d}\tau}\left(\|\boldsymbol{u}\|_\Omega^2\right)$$

is also bounded, as can be seen by the following arguments.

80) This can also be used for numerical purposes.

81) See, for example, R. Dautrey, J.-L. Lions: *Mathematical analysis and numerical methods for science and technology, vol. 2.* Berlin: Springer 1988, p. 126

Differentiate the differential equation in (5.26) with respect to τ. This leads in a first step that is analogous to (5.29) to

$$\frac{\mathrm{d}}{\mathrm{d}\tau}\left(\|\partial_\tau \boldsymbol{u}\|_\Omega^2 + \frac{1}{\beta^2}\|\partial_\tau p\|_\Omega^2\right) \leq -c\|\partial_\tau \boldsymbol{u}\|_\Omega^2 , \tag{5.31}$$

and hence to the boundedness of $\|\partial_\tau \boldsymbol{u}\|$. Because of

$$\left|\frac{\mathrm{d}}{\mathrm{d}\tau}\left(\|\boldsymbol{u}\|_\Omega^2\right)\right| = 2|(\boldsymbol{u}, \partial_\tau \boldsymbol{u})| \leq 2\|\boldsymbol{u}\|_\Omega \|\partial_\tau \boldsymbol{u}\|_\Omega ,$$

the existence of a constant ω with

$$\left|\frac{\mathrm{d}}{\mathrm{d}\tau}\left(\|\boldsymbol{u}\|_\Omega^2\right)\right| \leq \omega \tag{5.32}$$

follows. The inequalities (5.32) and (5.30) then imply

$$\lim_{\tau\to\infty} \|\boldsymbol{u}(\tau)\|_\Omega = 0 .$$

Otherwise there would be an $\varepsilon > 0$ and a number $\tau_i > \tilde{\tau}$ for every $\tilde{\tau} > 0$ with $\|\boldsymbol{u}(\tau_i)\|_\Omega^2 \geq \varepsilon$, and by means of the inequality

$$-\omega(\tau - \tau_i) \leq \|\boldsymbol{u}(\tau)\|_\Omega^2 - \|\boldsymbol{u}(\tau_i)\|_\Omega^2 \leq \omega(\tau - \tau_i) ,$$

which follows for every $\tau > \tau_i$ from (5.32),

$$\|\boldsymbol{u}(\tau)\|_\Omega^2 \geq \varepsilon - \omega(\tau - \tau_i)$$

would follow. Thus,

$$J_i := \int_{\tau_i}^{\tau_i+\frac{\varepsilon}{\omega}} \|\boldsymbol{u}(\tau)\|_\Omega^2 \,\mathrm{d}\tau \geq \frac{\varepsilon^2}{\omega} - \frac{\omega}{2}\frac{\varepsilon^2}{\omega^2} = \frac{\varepsilon^2}{2\omega}$$

would also result.

Now, after choosing a suitable subsequence of the sequence $\{J_i\}$, the intervals $\tau_i \leq \tau \leq \tau_i + \frac{\varepsilon}{\omega}$ $(i = 1, 2, \cdots)$ could be assumed to be disjunct, leading to

$$\sum_i J_i \leq \int_{\tau_1}^{\infty} \|\boldsymbol{u}(\tau)\|_\Omega^2 \,\mathrm{d}\tau < \infty .$$

This finally would contradict the unboundedness of the partial sums of $\sum_i J_i$.

5.4
Foundations of the Boundary Layer Theory and Flow Separation

Let us now look at the other extreme situation for the Navier–Stokes equations written in the dimensionless form (5.11), namely the case of large Reynolds numbers, which was studied by Prandtl in particular. Here, we again omit the tildes used in (5.11), and because the Reynolds numbers are large, we can expect the solutions to behave in relatively similar way to the solutions of the Euler equations.[82)]

Prandtl published his results in 1904 at the International Congress of Mathematicians in Heidelberg.[83)] Let us now discuss his ideas.

Assume that a solid is placed in a viscous flow. Let the no-slip condition (1.54) hold along the wall Γ of this solid, i.e., $\boldsymbol{u}_\Gamma = 0$, omit any influences of exterior forces, and assume that the flow was irrotational before the solid was placed in it.

Thus, at this moment the velocity $\boldsymbol{u}$ could be derived from a potential Φ which fulfills the potential equation $\Delta\Phi = 0$ because of the incompressibility of the fluid. This leads to the fact that there is no difference from the frictionless situation.

Prandtl restricted his considerations to a piece Γ of the surface of the solid which can be assumed to be (more or less) planar,[84)] and this plane was then used as the (x, y)-plane in an (x, y, z)-coordinate system. At a great distance from the plane, theoretically for $z \to \infty$, Prandtl expected the flow to retain its character as a potential flow, so that the Euler equations hold for large values of z.

The idea that the flow could perhaps be irrotational everywhere outside the solid immediately leads to a contradiction for $\boldsymbol{u} \not\equiv 0$, because the problem

$$\begin{aligned} \Delta\Phi &= 0 \quad \text{outside the solid} \\ \boldsymbol{u}_\Gamma &= 0 \quad \text{no-slip condition} \end{aligned}$$

only has the solution $\boldsymbol{u} \equiv 0$. We omit the proof and consider that a jump from the velocity $\boldsymbol{u} = 0$ along Γ to a potential flow of velocity $\boldsymbol{u} \neq 0$ outside the solid is also impossible, because the existence of the viscosity term does not allow discontinuities of this type.

Hence, the flow between the surface of the solid and the distant parts of the flow cannot be irrotational. The domain of this rotational flow is called the *boundary layer*. Here, the word *distant* has a rather vague meaning from the point of view of real applications. An obvious question is: what is the thickness δ of the boundary layer (i.e., what is distance at which the difference between the real flow and the

82) It took many years to prove mathematically that the Navier–Stokes solutions are close to the Euler solutions for small values of η: K. Nickel showed in Arch. Rat. Mech. Anal. **13** (1963) 1–14 that both solutions can, under certain conditions and for $\eta \to 0$, be asymptotically considered to be solutions of the Prandtl boundary layer equations (5.48) presented in this section. This also subsequently justified Prandtl's assumption regarding the small diameter of the boundary layer that we will use later.

83) In: Verh. d. III. Int. Math.-Kongr. Heidelberg 1904. Leipzig: Teubner 1905, pp. 484–494

84) W. Tollmien showed that all of these considerations also hold for curved surfaces provided that the curvature does not vary too much; see Handb. d. Exper.-Physik **IV**, I (1938) 248ff.

original potential flow can be neglected)? Of course, the answer to this depends on a convention. Often, a 1% difference between the velocities is an accepted value.

Prandtl assumed that the thickness δ was small compared with the length L used in order to write the Navier–Stokes equations in the dimensionless form (5.11), i.e.,

$$\delta \ll 1\,, \tag{5.33}$$

and he assumed that the transition from rotational flow to irrotational potential flow is continuous.

Let all of the derivatives occurring in the following considerations exist; let them be continuous; and let the dimensionless velocity $\boldsymbol{U}$ of the distant potential flow be parallel to Γ, i.e., parallel to the (x, y)-plane. Hence,

$$\boldsymbol{U} = \boldsymbol{U}(x, y, t) = \left(U(x, y, t), V(x, y, t), 0\right)^T .$$

Moreover, the components of $\boldsymbol{U}$ are assumed to be of the same order of magnitude as the velocity u_0 used to derive the dimensionless form (5.11) of the Navier–Stokes equations, so that U and V are of the order of magnitude 1. Because of the continuous transition from $\boldsymbol{u} = (u, v, w)^T$ to $\boldsymbol{U}$, the assumptions

$$u = \mathcal{O}(1) \quad \text{and} \quad v = \mathcal{O}(1) \quad \text{for} \quad z \longrightarrow \delta \tag{5.34}$$

with Landau symbol $\mathcal{O}$ are reasonable, and this will certainly also be true along all of the layers parallel to Γ, particularly for $x, y \longrightarrow 1$ where x and y are the dimensionless spatial variables in (5.11).

Concerning the spatial derivatives of the u- and v-components parallel to Γ, as well as their time derivatives, Prandtl expected that they would not vary very much within the boundary layer, i.e.,

$$\partial_t u, \partial_t v, \partial_x u, \partial_y u, \partial_{xx} u, \partial_{yy} u, \partial_x v, \partial_y v, \partial_{xx} v, \partial_{yy} v = \mathcal{O}(1)\,. \tag{5.35}$$

If the continuity equation $\operatorname{div} \boldsymbol{u} = 0$ is also taken into account, (5.35) yields a further result, namely

$$\partial_z w = \mathcal{O}(1) \quad \text{for} \quad z \longrightarrow \delta \quad \text{or for} \quad x, y \longrightarrow 1\,, \tag{5.36}$$

so that the no-slip condition leads to

$$|w(x, y, z, t)| = \left| \int_0^z \partial_z w(x, y, \zeta, t)\, \mathrm{d}\zeta \right| \le \|\partial_z w\|_\infty |z|\,,$$

i.e.,

$$w = \mathcal{O}(\delta)\,. \tag{5.37}$$

Combining assumption (5.33) with the no-slip condition, the mean value theorem leads inside the boundary layer to

$$u(x, y, z, t) \approx \partial_z u(x, y, z, t) \cdot z\,,$$

and therefore with (5.34) to

$$\partial_z u = \mathcal{O}\left(\frac{1}{\delta}\right) . \tag{5.38}$$

Analogously, the relations

$$\partial_z v = \mathcal{O}\left(\frac{1}{\delta}\right) , \quad \partial_{zz} u = \mathcal{O}\left(\frac{1}{\delta^2}\right) , \quad \partial_{zz} v = \mathcal{O}\left(\frac{1}{\delta^2}\right) \tag{5.39}$$

also hold.

Using the mean value theorem, we also obtain the statement

$$w(x, y, z, t) = w(0, y, z, t) + \partial_x w(\vartheta x, y, z, t) \cdot x, \quad 0 < \vartheta < 1 ,$$

and because x can grow to an order of magnitude of 1, (5.37) yields

$$\partial_x w = \mathcal{O}(\delta) \tag{5.40}$$

and analogously

$$\partial_y w, \partial_{xx} w, \partial_{yy} w = \mathcal{O}(\delta) , \quad \partial_t w = \mathcal{O}(\delta) . \tag{5.41}$$

Moreover, taking (5.36) into account, we additionally find

$$\partial_{zz} w = \mathcal{O}\left(\frac{1}{\delta}\right) . \tag{5.42}$$

Let us now write down the three components of (5.11) separately. We will write each of the terms occuring in the equations, and below them we will note down their particular orders of magnitude (as far as they are already known). Here we follow the representation used by H. Schlichting in the first edition of his book on boundary-layer theory:[85)]

$$-\nabla p = \begin{cases} \begin{matrix} \partial_t u & + & u & \partial_x u & + & v & \partial_y u & + & w & \partial_z u & -\frac{1}{\mathrm{Re}} & (\partial_{xx} u & + & \partial_{yy} u & + & \partial_{zz} u) \\ 1 & & 1 & 1 & & 1 & 1 & & \delta & \frac{1}{\delta} & & 1 & & 1 & & \frac{1}{\delta^2} \\ \\ \partial_t v & + & u & \partial_x v & + & v & \partial_y v & + & w & \partial_z v & -\frac{1}{\mathrm{Re}} & (\partial_{xx} v & + & \partial_{yy} v & + & \partial_{zz} v) \\ 1 & & 1 & 1 & & 1 & 1 & & \delta & \frac{1}{\delta} & & 1 & & 1 & & \frac{1}{\delta^2} \\ \\ \partial_t w & + & u & \partial_x w & + & v & \partial_y w & + & w & \partial_z w & -\frac{1}{\mathrm{Re}} & (\partial_{xx} w & + & \partial_{yy} w & + & \partial_{zz} w) \\ \delta & & 1 & \delta & & 1 & \delta & & \delta & 1 & & \delta & & \delta & & \frac{1}{\delta} \end{matrix} \end{cases} \tag{5.43}$$

85) H. Schlichting: *Grenzschichttheorie.* Karlsruhe: Braun 1951

Of course, the pressure gradient parallel to the plane is assumed to be bounded in all real applications, i.e.,

$$\partial_x p, \partial_y p = \mathcal{O}(1) \ .$$

However, this implies that the first two equations can only be valid with respect to the order of magnitude if

$$\mathrm{Re} = \mathcal{O}\left(\frac{1}{\delta^2}\right), \quad \text{i.e.} \quad \delta = \mathcal{O}\left(\frac{1}{\sqrt{\mathrm{Re}}}\right), \tag{5.44}$$

and, because of the third equation, this leads to

$$\partial_z p = \mathcal{O}(\delta) \ . \tag{5.45}$$

Equation (5.44) shows that Prandtl's assumption of a small thickness for the boundary layer for large Reynolds numbers does not lead to a contradiction.

It also shows that the boundary layer theory can *only* be accepted as an approximation for the Navier–Stokes equations in the case of large Reynolds numbers.

Because of

$$|p(x, y, z, t) - p(x, y, 0, t)| = \left| \int_0^z \partial_z p(x, y, \zeta, t) \, \mathrm{d}\zeta \right| \le |z| \, \| \, \partial_z p \, \|_\infty \ ,$$

the pressure differences in the z-direction within the boundary layer are of order δ^2, and are thus very small for large Reynolds numbers. We therefore put

$$\partial_z p \equiv 0 \ ,$$

so that, approximately, p does not depend on z within the boundary layer:

$$p = p(x, y, t) \ . \tag{5.46}$$

The pressure gradient in the boundary layer thus coincides with the well-known pressure gradient of the potential flow outside the boundary layer where the Euler equations hold. Using dimensionless variables and taking $W = 0$ into account, we therefore approximately obtain

$$\begin{aligned} \partial_t U + U\partial_x U + V\partial_y U &= -\partial_x p \\ \partial_t V + U\partial_x V + V\partial_y V &= -\partial_y p \end{aligned} \tag{5.47}$$

so that $\partial_x p$ and $\partial_y p$ are known quantities.

If all of the terms in (5.43) of order δ are now neglected compared with the terms of order 1, the system of three equations reduces to only two equations, and in dimensionless form these equations are:

$$\begin{aligned} \partial_t u + u\partial_x u + v\partial_y u + w\partial_z u - \nu\partial_{zz} u &= -\frac{1}{\varrho}\partial_x p \\ \partial_t v + u\partial_x v + v\partial_y v + w\partial_z v - \nu\partial_{zz} v &= -\frac{1}{\varrho}\partial_y p \end{aligned} \tag{5.48}$$

(5.48) are the *Prandtl boundary layer equations* with the incompressibility condition

$$\partial_x u + \partial_y v + \partial_z w = 0 \,. \tag{5.49}$$

The velocity component w remains in the system (5.48), but is now only a parameter which will later be fixed by (5.49) and by boundary conditions. Such boundary conditions may, for example, be:

$$u = v = w = 0 \quad \text{for} \quad z = 0$$
$$\begin{pmatrix} u \\ v \\ w \end{pmatrix} = \begin{pmatrix} U(x, y, t) \\ V(x, y, t) \\ 0 \end{pmatrix} \quad \text{for} \quad z \to \infty \,.$$

$z \to \infty$ and $\delta \ll L$ seem to be contradictory, but one must note that the boundary layer is infinitely thick from a theoretical point of view, while their effective thickness from a physical point of view is very small.

If the boundary layer is affected by sucking off or blowing up parts of it, the boundary conditions must model these effects, e.g., by replacing the equation $w = 0$ in (5.50) by

$$w = w_0(x, y, t) \quad \text{for} \quad z = 0 \,,$$

provided that the direction of the sucking or blowing coincides with the w-direction and that its flow velocity w_0 is of order $\mathcal{O}(\delta)$, i.e.,

$$w_0 = \mathcal{O}\left(\frac{1}{\sqrt{\text{Re}}}\right) . \tag{5.50}$$

Otherwise some of the terms that Prandtl expected to be very small can no longer be neglected.

As well as boundary conditions, initial conditions must also be formulated, and not just with respect to time in the case of instationary problems, but also with respect to space too in every case, because the boundary layer equations are parabolic. This parabolicity often includes simplifications when the construction of approximate analytic solutions of the Navier–Stokes equations is desired. However, Prandtl's theory can often make life easier numerically.

If the problem to be treated concerns an instationary flow, the initial task is to formulate initial conditions with respect to time, i.e., conditions of the type

$$\begin{aligned} u(x, y, z, 0) &= f_1(x, y, z) \\ v(x, y, z, 0) &= f_2(x, y, z) \end{aligned} \tag{5.51}$$

and in each case with respect to space too, i.e., to stipulate suitable initial profiles for the u- and v-components of the flow velocity.

There is no general consensus about what the initial profiles should look like, but it should be noted that the influence of the initial profiles decreases as the distance from the area where these profiles were prescribed increases. This follows from the parabolic situation.

These spatial initial conditions may take into consideration that a boundary layer comes into being only when it is a certain distance from the edges of the solid placed in the fluid, e.g., from the front edge of a wing:

$$\begin{aligned} u(x_0, y, z, t) &= g_{11}(y, z, t) \\ v(x_0, y, z, t) &= g_{12}(y, z, t) \end{aligned} \tag{5.52}$$

and

$$\begin{aligned} u(x, y_0, z, t) &= g_{21}(x, z, t) \\ v(x, y_0, z, t) &= g_{22}(x, z, t) \ . \end{aligned} \tag{5.53}$$

Of course, the conditions prescribed should go well together, e.g.,

$$f_1(x_0, y, z) = g_{11}(y, z, 0) \tag{5.54}$$

etc.

Many flow phenomenona that place along the surface of a solid can be quantitatively explained via the boundary layer concept. One of them is the separation of the boundary layer flow from a curved surface.

Here, we are going to describe this separation briefly and only qualitatively in the case of stationary 2-D flows.

Assume that an ideal fluid is flowing along a curved surface. A lateral flow with a main flow velocity $\boldsymbol{u} = (u_\infty, 0, 0)^{\mathrm{T}}$, $u_\infty > 0$ around a circular cylinder of infinite length whose axis coincides with the z-axis is a 2-D example. The velocity of the particles arriving at the surface of the cylinder increases due to Bernoulli's equation (1.31) because they initially move against decreasing pressure. The velocity then begins to decrease because the pressure grows again. Finally, the particles move behind the cylinder with their original speed (cf. Fig. 5.4).

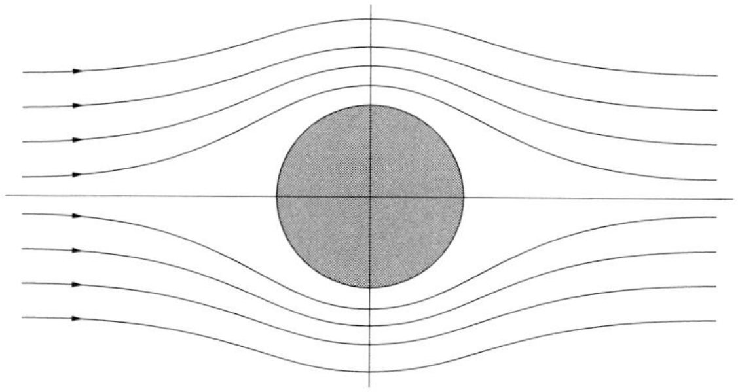

Fig. 5.4 Ideal flow around a cylinder.

In the case of a viscid flow, a certain amount of the kinetic energy of the particles will be transferred in the first phase to friction, so there will not be enough energy available in the second phase to return the particles to their former speed.

It can even happen that the particles stop and that the velocity reverses direction due to the increasing pressure; i.e., separation of the boundary layer takes place.

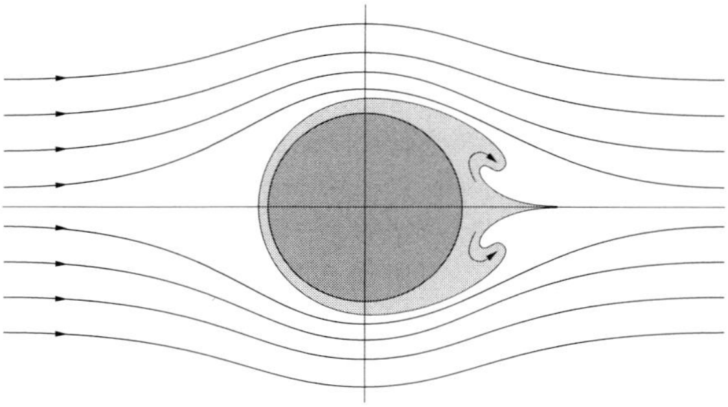

Fig. 5.5 Flow with separation.

In the 2-D situation, the point of separation at the curved surface, which is in the 2-D-dimensions a curve Γ, can be well described mathematically. For this purpose, let u be the particular component of the flow velocity parallel to Γ and let $\partial_n u$ be the normal derivative of this component. The separation point $P \in \Gamma$ is then obviously the point where this normal derivative vanishes (cf. Fig. 5.6):

$$\partial_n u(P) = 0 \,. \tag{5.55}$$

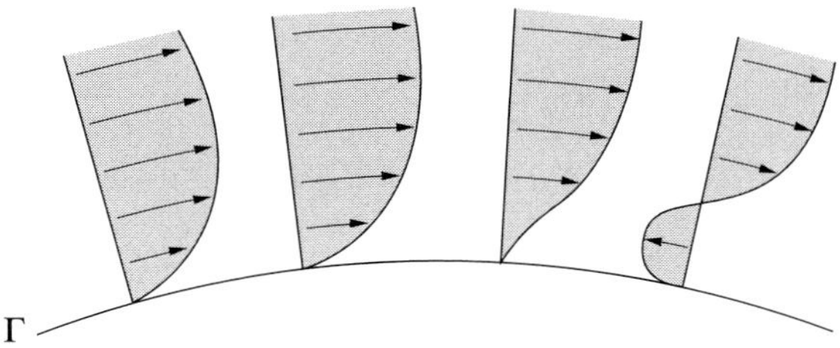

Fig. 5.6 Explanation for (5.55) in the case of 2-D boundary layer flow.

Beyond the point of separation, the flow loses its buoyancy effect. Hence, one tries to design a wing in such a way that the point of separation is situated as close as possible to the back edge of the wing, e.g., by sucking off the boundary layer.

It should be mentioned that there is no general mathematical definition of separation in the case of 3-D flows.

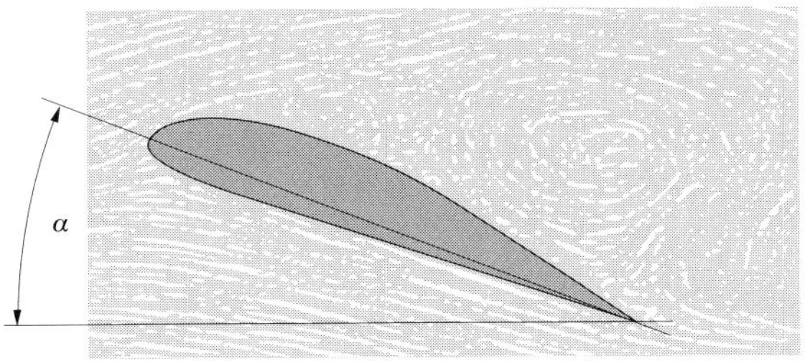

Fig. 5.7 Boundary layer separation on a wing.

5.5 Stability of Laminar Flows

The beginning of the transition of a flow from the laminar state to a turbulent one can be looked upon as a transition from stable behavior of the solution to the Navier–Stokes equations for this flow to instability.

In order to investigate this idea, for convenience, let us now consider a situation that is easy to survey, namely an incompressible 2-D flow and the knowledge of a stationary average laminar flow in the x-direction which has a velocity U that only depends on the second space variable y:

$$\boldsymbol{U} = \begin{pmatrix} U(y) \\ 0 \end{pmatrix} .$$

The pressure of this flow is denoted by $P(x, y)$, and $(\boldsymbol{U}, P)$ is assumed to fulfill the given incompressible Navier–Stokes problem[86] and to be one part of a superposition, the other part of which consists of infinitesimal and incompressible disturbances $(\tilde{\boldsymbol{u}}, \tilde{p})$ with

$$\tilde{\boldsymbol{u}} = \begin{pmatrix} \tilde{u}(x, y, t) \\ \tilde{v}(x, y, t) \end{pmatrix} \quad \text{and} \quad \tilde{p} = \tilde{p}(x, y, t) .$$

This second part is not necessarily a stationary flow and, besides $(\boldsymbol{U}, P)$, the complete flow

$$\boldsymbol{u} = \boldsymbol{U} + \tilde{\boldsymbol{u}} , \quad p = P + \tilde{p} \tag{5.56}$$

is also assumed to be a solution of the given Navier–Stokes problem.[87]

Because the disturbances are infinitesimal, quadratic terms can be neglected compared with linear terms, so that putting (5.56) into the Navier–Stokes equa-

86) It is absolutely problematic to suppose such an average flow to exist, but it helps to understand that turbulence is an instability phenomen.

87) This assumption is called the *closing*, and the treatment of the instability problem in this way is called the *Reynolds averaging*.

tions (5.10) and into the continuity equation then yields the relations

$$\begin{aligned}
&\partial_t \tilde{u} + U\partial_x \tilde{u} + \tilde{v}\frac{\mathrm{d}U}{\mathrm{d}y} + \frac{1}{\varrho}\partial_x P + \frac{1}{\varrho}\partial_x \tilde{p} = \nu\left(\frac{\mathrm{d}^2 U}{\mathrm{d}y^2} + \Delta\tilde{u}\right)\\
&\partial_t \tilde{v} + U\partial_x \tilde{v} \qquad\qquad + \frac{1}{\varrho}\partial_y P + \frac{1}{\varrho}\partial_y \tilde{p} = \nu\Delta\tilde{v}\\
&\qquad\qquad\qquad\qquad \partial_x \tilde{u} + \partial_y \tilde{v} = 0\,.
\end{aligned} \tag{5.57}$$

Because the average flow solves the Navier–Stokes equations on its own, (5.57) reduces to the linear equations

$$\begin{aligned}
&\partial_t \tilde{u} + U\partial_x \tilde{u} + \tilde{v}\frac{\mathrm{d}U}{\mathrm{d}y} + \frac{1}{\varrho}\partial_x \tilde{p} = \nu\Delta\tilde{u}\\
&\partial_t \tilde{v} + U\partial_x \tilde{v} \qquad\qquad + \frac{1}{\varrho}\partial_y \tilde{p} = \nu\Delta\tilde{v}\\
&\qquad\qquad \partial_x \tilde{u} + \partial_y \tilde{v} = 0\,.
\end{aligned} \tag{5.58}$$

If the first equation in (5.58) is differentiated with respect to y and the second one with respect to x, the pressure can be eliminated so that only one equation remains for u and v besides the continuity equation. The no-slip condition along surfaces of solids leads to additional boundary conditions.

Let us now suppose that the disturbances result from the superposition of waves propagating in the x-direction, where each of them fulfills (5.58). These waves can be expressed by

$$\tilde{\boldsymbol{u}} = \begin{pmatrix} \varphi(y) \\ \psi(y) \end{pmatrix} \mathrm{e}^{\mathrm{i}(\alpha x - \gamma t)} \tag{5.59}$$

and are called *Tollmien–Schlichting waves*. Here, $\alpha \in \mathbb{C}$, $\gamma \in \mathbb{C}$. The particular case $\alpha \in \mathbb{R}$ describes the investigation of the *temporal stability*, whereas an investigation of the case $\alpha \in \mathbb{C}$, $\gamma \in \mathbb{R}$ concerns the so-called *spatial stability*, which is not discussed here.

The continuity equation leads to

$$\varphi(y) = \frac{\mathrm{i}}{\alpha}\psi'(y)\,,$$

and the equation that remains after eliminating $\tilde{p}$ then yields

$$(U - \gamma)(\psi'' - \alpha^2\psi) - U''\psi = -\frac{\mathrm{i}\nu}{\alpha}(\psi^{(4)} - 2\alpha^2\psi'' + \alpha^4\psi)\,. \tag{5.60}$$

The linear homogeneous fourth-order ordinary differential equation (5.60) for ψ with complex coefficients is called the *Orr–Sommerfeld equation*. Boundary conditions are for instance the no-slip condition $\psi = 0$ for $y = 0$ and the disappearance of the disturbances for $y \to \infty$.

Thus, for every fixed $\alpha \in \mathbb{R}$, an eigenvalue problem arises where the complex *phase speed of perturbation* γ plays the role of the eigenvalue. The parameter ν [88] is considered to be already known from the laminar average flow. The laminar flow corresponds to the *trivial solution* $\psi \equiv 0$ of problem (5.60), and one asks for nontrivial solutions that bifurcate from it.

If the sign of the imaginary part of γ is negative, the disturbance decreases as time increases, and its amplitude does not depend on time if this imaginary part vanishes. However, an unstable situation occurs for

$$\operatorname{sgn}(\operatorname{Im}\gamma) > 0\,. \tag{5.61}$$

Under realistic boundary conditions for U, the Orr–Sommerfeld equation normally not be solved explicitly, though it is linear and homogeneous. Numerical procedures known from numerical bifurcation theory must be used.

A very thorough treatment of the Orr–Sommerfeld equation with boundary conditions corresponding to the particular situation of a half-plane flow along a wall was given by J.-R. Lahmann and M. Plum.[89]

The smallest value of the Reynolds number that gives validity to (5.61) is called the *critical Reynolds number* $\mathrm{Re}_{\mathrm{cr}}$, and is of interest because it is the smallest Reynolds number for which the flow can become unstable. The curve $\operatorname{sgn}(\operatorname{Im}\gamma) = 0$ obviously separates the area of instability in the (α, Re)-plane from the area of a stable laminar flow and is called the *curve of neutral stability*.

It should be noted that an unstable result from this linearized theory of disturbances does not necessarily mean that the real underlying flow can already be considered to be completely turbulent. One measure accepted by engineers is the so-called e^9*-rule*: the flow is actually a turbulent one if the small amplitude of the initial disturbance has increased by a factor $e^9 \approx 8100$.

5.6
Heated Real Gas Flows

The total energy E per unit volume, consisting of stored heat plus kinetic energy, is given by

$$E = \varrho c_v T + \frac{\varrho}{2}\|\boldsymbol{u}\|^2 \tag{5.62}$$

(ϱ is density, c_v is specific heat at constant volume, $\boldsymbol{u}$ is the velocity vector of the flow).

From Reynolds' transport theorem (1.3), we find

$$\frac{\partial}{\partial t}\int\limits_{W(t)} E\,\mathrm{d}W = \int\limits_{W(t)} \{E_t + \operatorname{div}(E\boldsymbol{u})\}\,\mathrm{d}W\,. \tag{5.63}$$

88) Or, if the problem is formulated in dimensionless form, $\frac{1}{\mathrm{Re}}$

89) J.-R. Lahmann, M. Plum: ZAMM **84** (2004) 188–204

On the other hand, the change in the total energy per unit time reads

$$\frac{\partial Q}{\partial t} + \int\limits_{W(t)} \left\langle \varrho \hat{\boldsymbol{k}}, \boldsymbol{u} \right\rangle \mathrm{d}W - \int\limits_{\partial W(t)} \langle p\boldsymbol{n}, \boldsymbol{u} \rangle \,\mathrm{d}o \tag{5.64}$$

(here, Q is the heat transferred into the gas volume, $\hat{\boldsymbol{k}}$ is the external force per unit mass, $\boldsymbol{n}$ is the outwardly directed unit vector on the surface $\partial W(t)$ of the volume $W(t)$ at the instant t.[90]).

In (5.64), $\frac{\partial Q}{\partial t}$ is represented by

$$\frac{\partial Q}{\partial t} = \kappa \int\limits_{\partial W(t)} \langle \nabla T, \boldsymbol{n} \rangle \,\mathrm{d}o + \int\limits_{W(t)} q \,\mathrm{d}W = \int\limits_{W(t)} \{\kappa \Delta T + q\} \,\mathrm{d}W \tag{5.65}$$

(q is the internal production of heat per unit volume, κ is the *thermal conductivity*).

Combination of the equations (5.62) to (5.65) leads to

$$\frac{\partial}{\partial t} \left(\varrho c_v T + \frac{\varrho}{2} \|\boldsymbol{u}\|^2 \right) + \operatorname{div} \left(\left\{ \varrho c_v T + \frac{\varrho}{2} \|\boldsymbol{u}\|^2 \right\} \boldsymbol{u} \right) = \kappa \Delta T + q + \langle \varrho \hat{\boldsymbol{k}}, \boldsymbol{u} \rangle - \operatorname{div}(p\boldsymbol{u}) \,,$$

i.e., to

$$\begin{aligned} &(\varrho c_v T)_t + \operatorname{div}(\{\varrho c_v T + p\}\boldsymbol{u}) + \varrho \langle \boldsymbol{u}_t, \boldsymbol{u} \rangle + \varrho_t \frac{\|\boldsymbol{u}\|^2}{2} \\ &+ \frac{\varrho}{2} \langle \boldsymbol{u}, \nabla(\|\boldsymbol{u}\|^2) \rangle + \frac{\|\boldsymbol{u}\|^2}{2} \operatorname{div}(\varrho \boldsymbol{u}) \\ &= \kappa \Delta T + q + \varrho \langle \hat{\mathbf{k}}, \boldsymbol{u} \rangle \,. \end{aligned} \tag{5.66}$$

Taking the continuity equation (1.4) into account as well as the Navier–Stokes equations (5.9), the last equation (5.65) leads to

$$\begin{aligned} &(\varrho c_v T)_t + \operatorname{div}(\{\varrho c_v T + p\}\boldsymbol{u}) - \varrho \left(\langle \boldsymbol{u}, \nabla \rangle \boldsymbol{u}, \boldsymbol{u} \right\rangle + (\lambda + \eta) \left\langle \nabla(\operatorname{div} \boldsymbol{u}), \boldsymbol{u} \right\rangle \\ &\quad + \eta \langle \boldsymbol{u}, \Delta \boldsymbol{u} \rangle - \langle \boldsymbol{u}, \nabla p \rangle + \frac{\varrho}{2} \langle \boldsymbol{u}, \nabla(\|\boldsymbol{u}\|^2) \rangle = \kappa \Delta T + q \,, \end{aligned} \tag{5.67}$$

i.e., to

$$(\varrho T)_t + \operatorname{div}(\varrho T \boldsymbol{u}) + \frac{1}{c_v} p \operatorname{div} \boldsymbol{u} = \frac{\kappa}{\varrho c_v} \left[\Delta(\varrho T) - T \Delta \varrho - 2 \langle \nabla \varrho, \nabla T \rangle \right] + \frac{1}{c_v} q - Vis \,, \tag{5.68}$$

where the viscosity terms *Vis* have the form

$$Vis = \frac{\lambda + \eta}{c_v} \langle \boldsymbol{u}, \nabla(\operatorname{div} \boldsymbol{u}) \rangle + \frac{\eta}{\varrho c_v} \langle \boldsymbol{u}, \Delta \boldsymbol{u} \rangle \,.$$

We shall later refer to the one-dimensional situation, where (5.67) becomes

$$(\varrho T)_t + (\varrho T u)_x + \frac{p}{c_v} u_x = \frac{\kappa}{c_v} T_{xx} + \frac{q}{c_v} - Vis \tag{5.69}$$

90) See Section 5.1

with

$$Vis = \frac{\hat{\eta}}{c_v} u u_{xx} ,$$

where we introduced the quantity $\hat{\eta}$, particularly

$$\hat{\eta} = \lambda + 2\eta$$

in the case of compressible gases.

Hence, the model of the heated gas flow consists of Eq. (5.67), of the continuity equation (1.4), the Navier–Stokes equations (5.9) and of an equation of state, e.g., for ideal gases

$$p = R\varrho T \tag{5.70}$$

where R is the *Boltzmann constant.*

5.7 Tunnel Fires

As mentioned at the end of Section 5.1, a particular situation occurs if there is a flow of heated gas in pipes where the heat is generated within the pipe. We say that a *tunnel fire* occurs if the heat is produced by burning objects, resulting in a mixture of air, smoke, etc., and high temperatures.

Tunnel fires have unfortunately occurred fairly often in recent years in rail and road tunnels, so, as well as fire brigades organizing experiments, it is in the public interest to study these events from a mathematical point of view.

Several authors have tried to describe this scenario using realistic mathematical models.[91] We refer here to the continuous model developed by I. Gasser and J. Struckmeier.[92]

From experience, the air or air/smoke mixture in the tunnel is homogeneous in the cross-sections of the tunnel, in the sense that there is no stratification of the flow with opposite flow velocity directions in different layers of the cross-section under consideration, at least if the velocities are not very small. Hence, we can take mean values of the quantities to be considered over the cross-section at a certain instant, which enables us to treat the problem as a one-dimensional one in space, where the dimensionless space variable $x (0 \leq x \leq 1)$ represents the longitudinal positions in the tunnel pipe. However, we must account for the fact that, in a 1D model, the reduction in pressure caused by the no-slip condition at the surface of the tunnel must be replaced by another suitable condition. We should also mention

91) See, e.g., G.B. Brandt, S.F. Jagger, C.J. Lea: Phil. Trans. R. Soc. London A **356** (1998) 2873–2906

92) I. Gasser, J. Struckmeier: Math. Meth. Appl. Sci. **25** (2002) 1231–1249

that, in this context, we will omit the heat loss at the tunnel surface, which is usually only small.

For convenience, we now introduce dimensionless, i.e., scaled parameters, variables and equations (as we did in the last part of Section 5.1). We do not distinguish between the unscaled and the scaled quantities in the text below. The reference values (e.g., ϱ_r, u_r) by which the unscaled values will be scaled (e.g., $\varrho_{\text{scaled}} = \varrho_{\text{unscaled}}/\varrho_r$, etc.) will be listed below.

We define the Reynolds number by

$$\text{Re} = \frac{\varrho_r u_r d}{\hat{\eta}} , \tag{5.71}$$

where d is a typical value for the cross-section of the tunnel.

The 1D model for the scaled density ϱ, the scaled velocity u, the scaled pressure p and the scaled temperature T then reads as follows.

In dimensionless form, the continuity equation (1.4) gives

$$\varrho_t + (\varrho u)_x = 0 , \tag{5.72}$$

while the Navier–Stokes equation (5.9) leads to

$$u_t + u u_x + \left(\frac{1}{\gamma M^2}\right) \frac{1}{\varrho} p_x = \left(\frac{1}{\text{Re}}\frac{d}{L}\right) \frac{1}{\varrho} u_{xx} + \hat{k} . \tag{5.73}$$

In the *unscaled* form, the exterior forces per unit mass are found to be

$$\hat{k} = -g \sin \alpha - f_1 \tag{5.74}$$

where $\alpha = \alpha(x)$ is the given slope profile of the tunnel, and g is the gravitational acceleration. In the 1D model, the (unscaled) term

$$f_1 = \frac{\xi}{d} \frac{u|u|}{2}$$

replaces the no-slip condition (1.54), i.e., the loss of force due friction at the tunnel surface, where the quadratic velocity term takes the turbulent character of the flow into consideration, and where ξ describes the surface of the tunnel and must be determined by experiment. Its order of magnitude is 10^{-2} to 10^{-1}.[93]

In the scaled version, we find:

$$\hat{k} = -\frac{gL}{u_r^2} \sin \alpha - \xi \frac{L}{d} \frac{u\,|u|}{2} . \tag{5.75}$$

In the dimensionless 1D description, (5.68) becomes

$$(\varrho T)_t + (\varrho u T)_x + (\gamma - 1) p u_x = \bar{\kappa} T_{xx} + \bar{q} q - M^2 \gamma (\gamma - 1) \bar{\eta} u u_{xx} \tag{5.76}$$

93) See, e.g., I. Gasser, R. Natalini: Quart. Appl. Math. 57 (1999) 269–282

Table 5.2 Reference values.

Quantity	Reference value	Typical magnitude
t	$t_r = L/u_r$	15 min
x	L	$10^3 - 10^4$ m
Cross-sectional width	d	10 m
u	u_r	$1\,\mathrm{m\,s^{-1}}$
ϱ	ϱ_r	$1.2\,\mathrm{kg\,m^{-3}} \cdot \mathbf{m^2}$
p	p_r	$1\,\mathrm{bar} = 10^5\,\mathrm{kg\,m^{-1}s^{-2}} \cdot \mathbf{m^2}$
f	f_r	$10\,\mathrm{m\,s^{-2}}$
T	$T_r = p_r/(\varrho_r R)$	300 K
q	$q_r = \gamma p_r u_r/((\gamma-1)L)$	$10^5 - 10^6\,\mathrm{W\,m^{-3}} \cdot \mathbf{m^2}$
R		$287\,\mathrm{m^2s^{-2}K^{-1}}$
c_p		$1005\,\mathrm{m^2s^{-2}K^{-1}}$
$\hat{\eta}$		$18 \times 10^{-6}\,\mathrm{kg\,m^{-1}s^{-1}} (= 18 \times 10^{-5}\,\mathrm{Poise}) \cdot \mathbf{m^2}$
κ		$25 \times 10^{-3}\,\mathrm{kg\,m\,s^{-3}K^{-1}} \cdot \mathbf{m^2}$

with

$$\bar{\eta} = \frac{1}{\mathrm{Re}}\frac{d}{L}\,, \quad \bar{\kappa} = \frac{1}{\gamma \mathrm{Pr}}\bar{\eta}\,, \quad \bar{q} = \frac{q_r L}{u_r p_r}(\gamma - 1)$$

where γ is from (1.9), $M^2 = \frac{\varrho_r u_r^2}{\gamma p_r}$, $\mathrm{Pr} = \frac{\hat{\eta} c_p}{\kappa}$ (cf. the last remark of Section 5.1), and Re is from (5.71).

Typical values for these parameters in our context are:

$$\gamma = 1.4\,, \quad M^2 = 8.6 \times 10^{-6}\,, \quad \mathrm{Re} = 6.7 \times 10^5\,, \quad \mathrm{Pr} = 0.72\,.$$

We must also take the equation of state into account. We assume that the gas in the tunnel behaves approximately like an ideal gas, so that (5.69) holds, which leads in the scaled form to

$$p = \varrho T\,. \tag{5.77}$$

Because the tunnel has a finite length L, boundary conditions will have to be described later, and the situation at the outbreak of the fire must be respected by the initial data.

Table 5.2 gives the reference values used here and their orders of magnitude. Without the bolded factor $\mathbf{m^2}$, they also fit the 3D case.

According to the Navier–Stokes equation there is a connection between the local pressure and the flow velocity at the point (x, t), and hence between the pressure and the Mach number.

If we introduce the quantity $\varepsilon := \gamma M^2$, which is on the order of 10^{-5} in the situation considered here, the expansion of p with respect to ε, i.e.,

$$p = p_0 + \varepsilon p_1 + \mathcal{O}(\varepsilon^2) \tag{5.78}$$

shows that a small Mach number asymptotic is reasonable.

p_0 is considered to be the mean outside pressure of the open tunnel region and is therefore assumed to be constant. It is also used as the reference value p_r in Table 5.2:

$$p_0 = p_r = \text{const} . \tag{5.79}$$

If one introduces (5.78) into (5.77) and neglects the small quantities of order $\mathcal{O}(\varepsilon)$,

$$p_0 = T\varrho \tag{5.80}$$

remains. Moreover, if we put (5.78) into (5.72), (5.73) and (5.76), these equations yield, via (5.80) and by omitting the $\mathcal{O}(\varepsilon)$ terms,

$$\varrho_t + (\varrho u)_x = 0 , \tag{5.81}$$

$$u_t + uu_x + \frac{1}{\varrho} p_{1x} = \frac{\bar{\eta}}{\varrho} u_{xx} + \hat{k} , \tag{5.82}$$

$$u_x = \bar{\kappa} \left(\frac{1}{\varrho} \right)_{xx} + \frac{1}{p_0} \bar{q} q . \tag{5.83}$$

Equation (5.83) shows that, under the assumptions of this section, the divergence of the flow cannot vanish. In other words, it cannot be treated as an incompressible flow.

From (5.76), we know that the quantities $\bar{\eta}$ and $\bar{\kappa}$ are very small, so that the viscous term in (5.82) and the heat conduction term in (5.83) can be omitted. If we also take (5.75) into account, we end up with the following system of equations:

$$\begin{aligned} \varrho_t + u\varrho_x &= -\varrho Q \\ u_t + uu_x + \frac{1}{\varrho} p_{1x} &= -\xi \frac{L}{d} \frac{u|u|}{2} - \frac{gL}{u_r{}^2} \sin\alpha \\ u_x &= Q \end{aligned} \tag{5.84}$$

with

$$Q = \frac{1}{\gamma p_0} \left(\frac{q_r L}{u_r p_0} \right) (\gamma - 1) q .$$

Initial data for the velocity $u(x,t)$ and the density $\varrho(x,t)$ must be prescribed.

If one replaces the last equation of the system (5.84) by the x-derived second one (and replaces u_x in the new equation by Q), and if this reformulated system together with realistic initial conditions is then associated with suitable Dirichlet boundary conditions for the pressure at the tunnel entrance and exit, by the homogeneous Neumann conditions

$$u_x(0,t) = u_x(1,t) = 0 , \quad \forall\, t > 0$$

for the velocity, and by the standard inflow boundary conditions

$$\varrho(0,t) = \varrho_0 \quad \text{if} \quad u(0,t) > 0 ; \quad \varrho(1,t) = \varrho_1 \quad \text{if} \quad u(1,t) < 0 , \quad \forall\, t > 0$$

for the density, realistic results can be obtained numerically.

Gasser and Struckmeier discretized system (5.84) by means of some classical explicit one-step finite difference schemes, and studied several situations. The results indicate that the model is able to perform real-time numerical experiments as well as to get qualitative insights into the flow conditions during a fire event. In this case, the dimensionless nature of the formulation makes it easy to look for these different situations.

We refer to the results of two of Gasser's and Struckmeier's numerical experiments: in both cases they considered tunnels of length $L = 4$ km and cross-sectional area 100 m^2. The fire occurs halfway between the ends of the tunnels and the effect of the heat sources amounts to 20 MW. The coefficient ξ in (5.75) is expected to be of order 0.1 in both cases, and the initial velocity to be 0 ms^{-1}.

Figure 5.8 shows the particular spatial distributions of the flow velocity, the pressure, the density and the temperature 20 min after a fire has broken out in a tunnel without a slope, where the pressures are 1010 mbar at the entrance and 1010.1 mbar at the exit.

In Fig. 5.9, we see the situation after 10 min in a tunnel with a slope of 3%, where the pressures are 1024.4 mbar at the entrance and 1010 mbar at the exit.

For the people running away from the location of the fire, it is important to know the flow direction of the hot gases. They will need to find this out beforehand, so that they will head in the correct direction along the tunnel (i.e., opposite to the

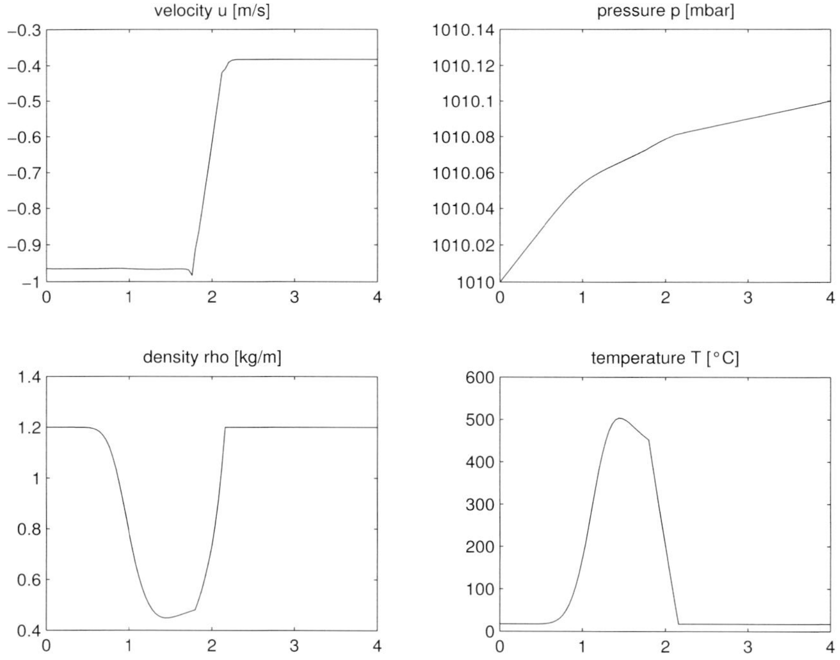

Fig. 5.8 Tunnel without a slope 20 min after a fire has broken out.

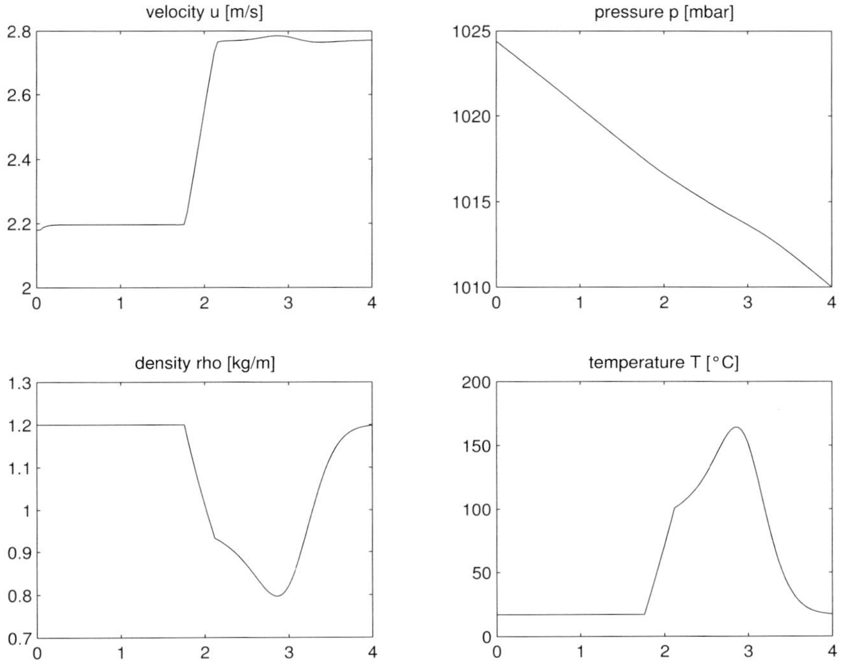

Fig. 5.9 Tunnel with a 3% slope 10 min after a fire has broken out.

direction of hot gas flow) in the event of fire. In other words, they need to learn how to behave in such a situation in order to avoid self-destruction through panic.

6
Proving the Existence of Entropy Solutions by Discretization Procedures

Though we have already studied some properties of weak solutions of conservation laws, we are yet to prove that this type of solution exists, particularly in the case of entropy solutions. The only exceptions were some scalar problems whose solutions could be specified explicitly. We are now going to address this issue.

6.1
Some Historical Remarks

Existence proofs for solutions of ordinary or partial differential equations are often structured in the following way. In a first step, the differential equation is discretized, as is normally done in numerical analysis in order to create numerical procedures. Thus, a finite-dimensional approximate version of the original problem that can be solved on a certain grid for each suitable step size parameter h is created. If the solution to the discrete problem is only given for a set of discrete points, it is extended to the inter-grid points by interpolation or by other techniques in such a way that it belongs to the particular function space that the solution or weak solution of the original problem is expected to belong to. This extension is called *reconstruction* (cf. the special case of (4.2)). Moreover, we assume that all of the computation that needs to be performed is realized theoretically without generating round-off errors.

If it can be shown in a second step that a sequence $\mathcal{L}$ of these reconstructed approximate solutions determined for each step size parameter h_j of a sequence $\{h_j\}$ of positive step-size parameters with $\lim_{j\to\infty} h_j = 0$ is compact in the function space under consideration, convergent subsequences of $\mathcal{L}$ exist. It can often be shown in a third step that, under certain conditions, the limits of those subsequences solve the original problem. Thus, solutions exist in this case, but different convergent subsequences can lead to different solutions of the original problem, because uniqueness has not yet been ensured.

A well-known example of an existence proof of this type is the particular proof of Peano's theorem on the existence of a solution to the initial value problem

$$y' = f(x, y)\ , \quad y(x_0) = y_0\ , \quad f \text{ is continuous}$$

Mathematical Models of Fluid Dynamics. R. Ansorge and T. Sonar

ISBN: 978-3-527-40774-3

where Euler's polygon method is used in order to construct approximate solutions in the space of functions that are continuous over a compact interval and equipped with the maximum norm.

Another example is the existence proof of Courant, Friedrichs and Lewy, in which this type of proof was applied for the first time to initial value problems for systems of quasilinear hyperbolic partial differential equations. The proof was published in their famous paper of 1928,[94] and it was in this paper that a condition concerning the step ratio was formulated in order to ensure the numerical stability of the numerical procedure, later called the CFL condition. This condition ensures that the region affecting the solution at a point Q where the solution is expected to exist is covered by the region that affects the approximate solution at Q, and this condition is extremely important when treating conservation laws by means of explicit numerical methods.[95]

Of course, because Courant, Friedrichs and Lewy asked for smooth solutions, they could only provide the conditions for local existence.

Under certain restrictions on the initial functions, J. Glimm provided an existence theorem for global weak solutions of conservation laws in a paper that also became famous,[96] and one of his tools was also a certain finite difference procedure.

We are now going to study a general theory on discretizations of partial differential equations that can also help to prove existence theorems for conservation laws such as those that occur in fluid dynamics.[97]

6.2 Reduction to Properties of Operator Sequences

In 1948, Kantorowitch published his important paper on the connections between numerical mathematics and functional analysis.[98]

Particularly focusing on linear partial evolution equations, in 1956 Lax and Richtmyer provided a functional analytic framework for the construction and investigation of finite difference methods[99] (cf. Section 7.4). Their theory was later generalized in respect to various aspects.

Inspired by the Lax–Richtmyer theory as well as the particular question of the discretization of differential equations, in 1972 Stummel published his general theory on discretization algorithms.[100] Though it was already rather general, it was still influenced by the linear structures underlying the given general problem. This theory was later also generalized in different ways and applied to several types of

94) R. Courant, K.-O. Friedrichs, H. Lewy: Math. Ann. **100** (1928) 32–74

95) For more details, see Section 7.3

96) J. Glimm: Comm. Pure Appl. Math. **18** (1965) 697–715

97) R. Ansorge: ZAMM **73** (1993) 239–253

98) L.W. Kantorowitch: Uspeki Mat. **3** (1948) 89–185

99) P.D. Lax, R.D. Richtmyer: Comm. Pure Appl. Math. **9** (1956) 267–293

100) F. Stummel in: J. Miller (ed.): *Proceedings of the Conference on Numerical Analysis Dublin 1972*. London: Academic 1973, pp. 285–310

problems, such as ordinary or partial differential equations, integral equations, optimization problems, etc. This theory is now called the *functional analysis of discretization algorithms* (Vainikko). However, weak solutions of differential equations or problems whose solutions are not necessarily unique had not been included in this theory.

Let us therefore, in a first step, try to describe how discretization techniques for problems of this type are embedded into the general theory, and to look for problems given in the following general form.

Let X, Y, X_n, Y_n be topological, usually metric spaces. We now ask for an element $v \in X$ which fulfills

$$Av = a\,, \quad A\colon\ X \to Y \quad \text{is a given operator}\,. \tag{6.1}$$

Let the numerical procedure constructed in order to solve this problem also be described in a general form by

$$A_n v_n = a_n\,, \quad A_n\colon\ X_n \to Y_n\,,\ n \in \mathbb{N}\,, \tag{6.2}$$

where $\{a_n | a_n \in Y_n,\ n \in \mathbb{N}\}$ is a given sequence approximating $a \in Y$ in a sense that is still to be specified.

Neither $X_n \subset X_{n+1}$ nor $X_n \subset X$ and neither $Y_n \subset Y_{n+1}$ nor $Y_n \subset Y$ $(n = 1, 2, \ldots)$ are assumed to hold.

Assume either that there are mappings

$$p_n : X \to X_n\,, \quad q_n : Y \to Y_n$$

$$\begin{array}{ccc} X & \xrightarrow{\ A\ } & Y \\ {\scriptstyle p_n}\downarrow & & \downarrow{\scriptstyle q_n} \\ X_n & \xrightarrow{\ A_n\ } & Y_n \end{array}$$

or that there is an embedding of the numerical procedure, just for theoretical purposes, into the framework of the original problem. This can for instance be done in the sense of

$$X_n \subset X\,, \quad Y_n = Y \quad (n = 1, 2, \ldots)\,,$$

but not necessarily

$$X_n \subset X_{n+1} \quad (n = 1, 2, \ldots)\,.$$

Just for convenience, let us suppose that such an imbedding is realized, but the original problem does not necessarily need to be linear, and its solutions are allowed to be weak in a sense that is still to be specified.

Let the original problem be described by

$$\tilde{A}v = \tilde{a}\,, \tag{6.3}$$

with $\tilde{A}\colon\ \tilde{X} \to Y$, $\tilde{X} \subset X$, $\tilde{a} \in Y$ a given right hand side.

It is supposed that the weak formulation of (6.3), which will then be the actual problem to solve, is described in the following way.

Let J be an index set, let $\{a(\Phi), \Phi \in J\} \subset Y$ be a given set, and let $\{A(\Phi), \Phi \in J\}$ be a set of operators with a joint domain $D \subset X$, $D \to Y$.

We ask for an element $v \in X$ with

$$A(\Phi)v = a(\Phi) \quad \forall \Phi \in J\,. \tag{6.4}$$

Definition

The elements of

$$\Theta := \{v \in X | v \ \text{ solves (6.4)}\}$$

are called *weak solutions* of (6.3) if the following implications hold:

If v solves (6.3), then $v \in \Theta$; if $v \in \tilde{X} \cap \Theta$, it also solves (6.3).

The elements $\Phi \in J$ are called *test elements.*

Also, let the numerical method be described in a rather general form by

$$\hat{A}_n v_n = \hat{a}_n \quad (n = 1, 2, \ldots) \tag{6.5}$$

where $\hat{A}_n$: $X_n \to Z$ $(n = 1, 2, \ldots)$, with a certain topological space Z. Here, $\{\hat{a}_n\} \subset Z$ is a given sequence that is compact in Z.

Assume

$$\Theta_n := \{v_n \in X_n | v_n \text{ solves (6.5)}\} \neq \emptyset \quad (n = 1, 2, \ldots)\,,$$

but that every Θ_n may contain more than one element. The solutions v_n of (6.5) are called *approximate solutions* of (6.4).

It is assumed that numerical method (6.5) does not depend on test elements, because computers cannot understand what test elements are. However, we also suppose that this method can be formulated weakly via a sequence of operators

$$\{A_n(\Phi) | \Phi \in J\,,\ n \in \mathbb{N}\} \ \text{with}\ A_n(\Phi):\ X_n \to Y_n\,,$$

and a sequence $\{a_n(\Phi)\}$ with

$$\lim_{n\to\infty} a_n(\Phi) = a(\Phi) \quad \text{for every fixed } \Phi \in J\,,$$

so that

$$A_n(\Phi)v_n = a_n(\Phi) \quad \forall \Phi \in J\,,\ \forall v_n \in \Theta_n \tag{6.6}$$

holds.

Equation (6.6) is called a *weak formulation of the numerical method.*

Some concepts help to prove certain convergence results.

Definition

A pair $[\{C_n\}, C]$ consisting of an operator sequence $\{C_n\}$ and an operator C is said to be *asymptotically closed* if the implication

$$v_n \to v \quad \wedge \quad C_n v_n \to z \quad \Rightarrow \quad Cv = z$$

holds.

Definition

An operator sequence $\{C_n\}$ is said to be *asymptotically regular* if the implication

$$\{C_n v_n\} \text{ compact in } Y \;\Rightarrow\; \{v_n\} \text{ compact in } X$$

holds.

Definition

Method (6.5) is called *convergent* if set convergence

$$\Theta_n \to \Theta$$

is ensured in the following sense:
$\{\Theta_n\}$ is *discretely compact*, i.e., every sequence $\{v_n | v_n \in \Theta_n;\ n = 1, 2, \ldots\}$ is compact in X. If v is a limit of a convergent subsequence, $v \in \Theta$ follows.

Remark

$\Theta_n \to \Theta$ implies $\Theta \notin \emptyset$. In other words, there is a solution of (6.4), i.e., a weak solution of (6.3).[101]

6.3 Convergence Theorems

We can now state a convergence theorem.

Theorem 6.1

(i) Let $[\{A_n(\Phi)\}, A(\Phi)]$ be asymptotically closed for every fixed $\Phi \in J$, and let
(ii) $\{\hat{A}_n\}$ be asymptotically regular.

Then:

$$\Theta_n \to \Theta\,.$$

101) A proof of $\Theta_n \to \Theta$ is therefore also an existence proof.

Proof:

$\{\hat{a}_n\}$ compact in $Z \Rightarrow \left\{\hat{A}_n v_n | v_n \in \Theta_n\right\}$ compact in Z, and this holds for every sequence $\{v_n | v_n \in \Theta_n\}$.

Thus, each of the sequences $\{v_n\}$ is compact in X because of assumption (ii), i.e.,

$$\exists \{v_n | v_n \in \Theta_n, n \in \mathbb{N}' \subset \mathbb{N}\}, \quad \exists v \in X \quad \text{with } v_n \to v \ (n \in \mathbb{N}') .$$

Here, $\mathbb{N}'$ and v are independent of Φ.

However, because v_n also fulfills $A_n(\Phi)v_n = a_n(\Phi)$, and because of

$$a_n(\Phi) \to a(\Phi) \quad \forall \Phi \in J ,$$

supposition (i) yields $A(\Phi)v = a(\Phi)$, i.e., $v \in \Theta$.

If one is only interested in a particular problem, i.e., in a particular right hand side of equation (6.3), assumption (ii) can obviously be replaced by the weaker requirement

(ii*) $\{\Theta_n\}$ discretely compact.

If the uniqueness of a weak solution is not guaranteed, sometimes additional restrictions are formulated in order to select from among the set of weak solutions the particular one that is relevant from the user's point of view. Also, as in the case of scalar conservation laws, this restriction often takes the form of an inequality that the relevant solution must additionally fulfill. Therefore, we will generally call such a restriction an *entropy condition*, and we formulate this condition in a general setting as

$$B(\hat{\Phi})v \leq 0 \quad \forall \hat{\Phi} \in \hat{J} , \tag{6.7}$$

where $\left\{B(\hat{\Phi}) | \hat{\Phi} \in \hat{J}\right\}$ denotes a set of nonlinear continuous functionals mapping X into $\mathbb{R}$, and where $\hat{J}$ is a certain index set which may differ from J.

We suppose that there is a uniqueness theorem available:[102)]

There is at most *one solution of* (6.4) *that also fulfills* (6.7);
this is termed the entropy solution.

If the original problem is accompanied by an entropy condition in order to characterize the relevant solution, it seems reasonable to also discretize this condition by

$$B_n(\hat{\Phi})v_n \leq 0 \quad \forall \hat{\Phi} \in \hat{J} \quad (n = 1, 2, \ldots) , \tag{6.8}$$

and to accept only approximate solutions $v_n \in \Theta_n$ which additionally fulfill this discrete entropy version. In this case, the functionals $B_n(\hat{\Phi})$ are also assumed to be continuous.

102) Theorem 3.3 is an example.

We suppose that

$$\{(6.5), (6.8)\} \text{ has } \textit{at least} \text{ one solution } \hat{v}_n \in \Theta_n \text{ for every fixed } n \in \mathbb{N}.$$

(6.8) is called a sequence of *discrete entropy conditions*, but we do not expect a *discrete entropy solution*

$$\hat{v}_n \quad (n \in \mathbb{N} \text{ arbitrary})$$

to necessarily be unique.

Definition

An operator sequence $\{C_n\}$ is said to be *continuously convergent* to an operator C if the implication

$$v_n \to v \quad \Rightarrow \quad C_n v_n \to Cv$$

holds.[103)104)] We then write

$$C_n \xrightarrow{c} C\,.$$

We can now supplement Theorem 6.1 with another important statement:

Theorem 6.2

Let

$$\hat{\Theta}_n := \{\hat{v}_n \in \Theta_n | \hat{v}_n \text{ solves } (6.8)\} \neq \emptyset\,,$$

and let

$$B_n(\hat{\Phi}) \xrightarrow{c} B(\hat{\Phi}) \quad \text{be fulfilled for every fixed } \hat{\Phi} \in \hat{J}\,.$$

Moreover, assume that the suppositions of Theorem 6.1 are fulfilled.
Then there is an entropy solution $\hat{v}$ of $\{(6.4),(6.7)\}$, and

$$\hat{\Theta}_n \to \{\hat{v}\}\,.$$

Proof: Because of the validity of Theorem 6.1, i.e., because of $\Theta_n \to \Theta$, there is a subsequence

$$\left\{\hat{v}_n | \hat{v}_n \in \hat{\Theta}_n,\ n \in \mathbb{N}' \subset \mathbb{N}\right\}$$

and a weak solution $v \in \Theta$ with $\{\hat{v}_n\} \to v$, $n \in \mathbb{N}'$.

103) P. du Bois-Reymond: Sitz.-Berichte, Preussische Akad. d. Wiss. (1886) 359–360
104) R. Courant: Göttinger Nachr. (1914) 101–109

The *continuous convergence* therefore yields $B_n(\hat{\Phi})\hat{v}_n \longrightarrow B(\hat{\Phi})v$.

From (6.8), $B(\hat{\Phi})v \leq 0$ follows. Consequently, v is an entropy solution, and because the entropy solution was assumed to be unique, v coincides with this entropy solution, i.e., $v = \hat{v}$, and so both the subsequence and the full sequence converge to $\hat{v}$:

$$\lim_{n \to \infty} \hat{v}_n = \hat{v} .$$

The conditions allowing for the property of *continuous convergence* on metric spaces follow from a theorem provided by W. Rinow:[105)106)]

Theorem 6.3

If the operators C, C_n $(n = 1, 2, \ldots)$ mapping a metric space X into a metric space Y are individually continuous, then $C_n \stackrel{c}{\longrightarrow} C$ is equivalent to the simultaneous validity of the two statements:

(i) $\{C_n\}$ is a sequence of equicontinuous operators

(ii) $C_n \longrightarrow C$ holds pointwise on a subset $\tilde{X} \underset{\text{dense}}{\subset} X$

6.4 Example

As an example, we treat the scalar version of problem (2.1). Thus, let $x \in \mathbb{R}$ be a spatial variable and $t \geq 0$ a time variable, and let Ω again be the upper (x, t) half-plane.

We ask for solutions $v : \Omega \longrightarrow \mathbb{R}$ of the scalar conservation law

$$\begin{aligned} \partial_t v + \partial_x f(v) &= 0, \ t > 0 \\ v(x, 0) &= v_0(x) \text{ with } v_0 \in BV(\mathbb{R}) \cap L_\infty(\mathbb{R}) , \end{aligned} \tag{6.9}$$

where the flux $f \in C^1$ is strictly convex and has its minimum (assumed to be zero) at zero: $f(0) = 0$. Let

$$|f'|_\infty^* := \max\left\{|f'(v)|, |v| \leq \|v_0\|_{L_\infty}\right\} < \infty . \tag{6.10}$$

Because of possible discontinuities, we are interested in weak global solutions $v \in X := L_1^{\text{loc}}$ of

$$\begin{aligned} A(\Phi)v := &- \int_\Omega\int \{v\partial_t\Phi + f(v)\partial_x\Phi\} \, \mathrm{d}x \, \mathrm{d}t \\ &- \int_{-\infty}^{\infty} v_0(x)\Phi(x, 0) \, \mathrm{d}x = 0 \quad \forall \, \Phi \in J := C_0^1(\Omega) . \end{aligned} \tag{6.11}$$

105) W. Rinow: *Die Innere Geometrie der Metrischen Räume*. Berlin: Springer 1961, 78ff.

106) See also Section 7.9

Let

$$\tilde{S} = \tilde{S}(v, c) , \quad \mathbb{R} \times \mathbb{R} \to \mathbb{R}$$

be a one-parameter set of so-called *entropy functions* that are continuously differentiable for every fixed $c \in \mathbb{R}$ with respect to v, possibly with an exception at $v = c$.

$\tilde{F} = \tilde{F}(v, c)(\mathbb{R} \times \mathbb{R} \to \mathbb{R})$ is called the *entropy flux* if

$$\partial_t \tilde{S}(v(x, t), c) + \partial_x \tilde{F}(u(x, t), c) = 0 \quad \text{on } \Omega'$$

holds for *every* fixed $c \in \mathbb{R}$ and for *every* pair (v, Ω') for which v is a weak solution on Ω but even a smooth C^1-solution on $\Omega' \subset \Omega$.

If a solution v which is weak on Ω additionally fulfills the inequality

$$\partial_t \tilde{S}(v(x, t), c) + \partial_x \tilde{F}(v(x, t), c) \le 0 \quad \forall c \in \mathbb{R}$$

on Ω in a weak sense, i.e.,

$$-\int_{\Omega}\int \{\partial_t \Phi(x, t) \tilde{S}(v(x, t), c) + \partial_x \Phi(x, t)\, \tilde{F}(v(x, t), c)\}\, dx\, dt - \int_{-\infty}^{\infty} \Phi(x, 0) \tilde{S}(v_0(x), c)\, dx \le 0 \tag{6.12}$$

$$\forall c \in \mathbb{R} , \quad \forall \textit{ nonnegative } \Phi \in J ,$$

then v is called the *entropy solution*, denoted by $\hat{v}$.

Thus, test elements in this example are the pairs

$$\hat{\Phi} := (\Phi, c) \text{ with } \hat{\Phi} \in \hat{J} := \{(\Phi, c) | \Phi \in J , \ \Phi \ge 0 , \ c \in \mathbb{R}\} ,$$

and the integral inequality can therefore be written as

$$B(\hat{\Phi})v \le 0 \quad \forall \hat{\Phi} \in \hat{J} \tag{6.13}$$

in accordance with the general theory. Particular choices of $\tilde{S}$ in the literature are

$\tilde{S} = \tilde{S}(v)$,	$\tilde{S}$ independent of c but strictly convex, $\hat{\Phi} = \{\Phi \in J \| \Phi \ge 0\} \subset J$ (Lax; cf. (3.11)) ,
$\tilde{S}(v, c) = \|v - c\|$	(Kruzkov; cf. (3.44)) ,
$\tilde{S}(v, c) = \begin{cases} v - c , & v \ge c \\ 0 , & v < c \end{cases}$	(Harten, Hyman, Lax)[107]
etc.	

Each of these choices then implies a particular choice of $\tilde{F}$ according to (3.13).

107) A. Harten, J.M. Hyman, P.D. Lax: Comm. Pure Appl. Math. **29** (1976) 297–322

Let the numerical procedure used in the general setting be directed into a more concrete form using a $(2k+1)$-point finite difference scheme:

$$\frac{v_j^{\nu+1} - v_j^\nu}{\Delta t} + \frac{g_{j+\frac{1}{2}}^\nu - g_{j-\frac{1}{2}}^\nu}{\Delta x} = 0 , \tag{6.14}$$

where ν counts the number of time steps and j the number of spatial steps, starting from

$$v_j^0 = \frac{1}{\Delta x} \int_{(j-1/2)\Delta x}^{(j+1/2)\Delta x} v_0(x)\, \mathrm{d}x .$$

Here, the *numerical flux*

$$g_{j+\frac{1}{2}}^\nu := g\left(v_{j-k+1}^\nu, v_{j-k+2}^\nu, \ldots, v_j^\nu, v_{j+1}^\nu, \ldots, v_{j+k}^\nu\right)$$

is assumed to be Lipschitz continuous and *consistent* with the original problem, i.e.,

$$g(w, w, \ldots, w) = f(w) , \quad \forall\, w \in \mathbb{R} . \tag{6.15}$$

Methods of this form are called *methods in conservation form.*

We restrict the choices of the step sizes by requiring that the CFL condition, introduced at the beginning of this chapter, must be respected. For the problem considered here, this means that

$$0 < \lambda := \frac{\Delta t}{\Delta x} = \text{const} \le \frac{1}{|f'|_\infty^*} . \tag{6.16}$$

Let $\Delta x = O(\frac{1}{n})$ $(n \in \mathbb{N})$, and put

$$v_n(x,t) = v_j^\nu \quad \text{for} \quad \begin{cases} \left(j-\frac{1}{2}\right)\Delta x \le & x < \left(j+\frac{1}{2}\right)\Delta x \\ \nu\,\Delta t \le & t < (\nu+1)\Delta t \end{cases}$$

$$(j = 0, \pm 1, \pm 2, \ldots) , \quad (\nu = 1, 2, \ldots) .$$

If X_n is then chosen as the space of functions defined on Ω and constant on these rectangles, the finite difference scheme can be formulated as

$$\hat{A}_n\, v_n = 0 \quad (n = 1, 2, \ldots) \tag{6.17}$$

with operators $\hat{A}_n\colon X_n \to Z := \mathbb{R}$.

Multiplication of (6.14) by test functions $\Phi \in J$ and integration leads to

$$A_n(\Phi) v_n = \tag{6.18}$$

$$\sum_{\nu=0}^{\infty} \sum_j \int_{\nu\Delta t}^{(\nu+1)\Delta t} \int_{(j-\frac{1}{2})\Delta x}^{(j+\frac{1}{2})\Delta x} \Phi(x,t) \left[v_j^{\nu+1} - v_j^\nu + \lambda\left(g_{j+\frac{1}{2}}^\nu - g_{j-\frac{1}{2}}^\nu\right)\right] \mathrm{d}x\, \mathrm{d}t = 0 .$$

This describes the operators $A_n(\Phi)$ as well as $a_n(\Phi)$ on the right hand side, which equal zero in this context.

Using the abbreviation

$$\tilde{S}_j^\nu(c) = \tilde{S}\left(v_j^\nu, c\right) ,$$

v_n is said to be a *discrete entropy solution*, denoted by $\hat{v}_n$, if

$$\frac{\tilde{S}_j^{\nu+1}(c) - \tilde{S}_j^\nu(c)}{\Delta t} + \frac{G_{j+\frac{1}{2}}^\nu(c) - G_{j-\frac{1}{2}}^\nu(c)}{\Delta x} \le 0 \quad \forall c \in \mathbb{R} , \tag{6.19}$$

where

$$G_{j+\frac{1}{2}}^\nu(c) := G(v_{j-k+1}^\nu, \dots, v_j^\nu, v_{j+1}^\nu, \dots, v_{j+k}^\nu, c)$$

denotes for every fixed c a Lipschitz-continuous *numerical entropy flux* that is still to be specified and is consistent with the entropy flux $\tilde{F}$ in the sense of

$$G(w, w, \dots, w, c) = \tilde{F}(w, c) \quad \forall\, (w, c) \in \mathbb{R}^2 . \tag{6.20}$$

For $k = 1$ (three-point case), the *flux splitting* choice

$$G(\alpha, \beta, c) = \tilde{F}_+(\alpha, c) + \tilde{F}_-(\beta, c) \tag{6.21}$$

with

$$\tilde{F}_+(\alpha, c) := \begin{cases} \tilde{F}(\alpha, c) , & \alpha \ge 0 \\ 0 , & \alpha < 0 \end{cases} , \qquad \tilde{F}_-(\beta, c) := \begin{cases} 0 , & \beta \ge 0 \\ \tilde{F}(\beta, c) , & \beta < 0 \end{cases}$$

is one of several possible choices of G, where $\tilde{F}$ is known as soon as $\tilde{S}$ is chosen.

The weak formulation of the discrete form of the entropy condition needed to use Theorem 6.2 arises from the multiplication of (6.19) by nonnegative test functions $\Phi \in J$ and the following integration:

$$\begin{aligned} B_n(\hat{\Phi})v_n := \int\!\!\int_\Omega \Phi(x, t) \Bigg\{ & \frac{1}{\Delta t} \left[\tilde{S}(v_n(x, t + \Delta t), c) - \tilde{S}(v_n(x, t), c)\right] \\ & + \frac{1}{\Delta x} \Big[G(v_n(x, t), v_n(x + \Delta x, t), c) \\ & - G(v_n(x - \Delta x, t), v_n(x, t), c)\Big] \Bigg\} \,\mathrm{d}x\, \mathrm{d}t \le 0 , \\ & \forall\, \hat{\Phi} \in \hat{J} , \end{aligned} \tag{6.22}$$

i.e.,

$$\begin{aligned} B_n(\hat{\Phi})\, v_n = & - \int_{\Delta t}^{\infty} \int_{-\infty}^{\infty} \frac{\Phi(x, t) - \Phi(x, t - \Delta t)}{\Delta t} \tilde{S}(v_n(x, t), c) \,\mathrm{d}x\, \mathrm{d}t \\ & - \frac{1}{\Delta t} \int_0^{\Delta t} \sum_i \int_{x_i - \Delta x/2}^{x_i + \Delta x/2} \Phi(x, t) \,\mathrm{d}x\, \tilde{S}\left(v_i^0, c\right) \,\mathrm{d}t \\ - \int\!\!\int_\Omega & \frac{\Phi(x + \frac{\Delta x}{2}, t) - \Phi(x - \frac{\Delta x}{2}, t)}{\Delta x} G\left(v_n\left(x - \tfrac{\Delta x}{2}, t\right), v_n\left(x + \tfrac{\Delta x}{2}, t\right), c\right) \mathrm{d}x\, \mathrm{d}t \\ & \le 0 , \quad \forall\, \hat{\Phi} \in \hat{J} . \end{aligned} \tag{6.23}$$

Thus, B and B_n from (6.7) and (6.8), respectively, are placed in a concrete form provided that a three-point scheme is being considered. If, for instance, the Harten–Hyman–Lax choice of $\tilde{S}$ is taken into account, and G is chosen according to (6.21), then $B(\hat{\Phi})$ in (6.12) and (6.13) is continuous for every fixed Φ. The functionals $B_n(\Phi)$ $(n = 1, 2, \ldots)$ in (6.22) are equicontinuous, and the pointwise convergence $B_n(\hat{\Phi}) \longrightarrow B(\hat{\Phi})$ can easily be demonstrated provided that the joint domain of these functionals is restricted to elements $v \in X^*$; here, X^* is the space of functions $v \in X$ restricted to the domain $\text{supp}(\Phi)$, which is equipped with the L_1 metric on X^*.

Hence, the assumptions (i) and (ii) of Rinow's theorem are fulfilled, i.e.,

$$B_n(\Phi) \xrightarrow{c} B(\Phi) ,$$

and Theorem 6.2 applies as far as $\hat{\Theta}_n \neq \emptyset$.[108]

The explicit[109] *Engquist–Osher three-point flux splitting scheme,*[110] characterized by (6.14) with

$$g^{\nu}_{j+\frac{1}{2}} = g\left(v^{\nu}_j, v^{\nu}_{j+1}\right) := f_-\left(v^{\nu}_{j+1}\right) + f_+\left(v^{\nu}_j\right) \tag{6.24}$$

where

$$f_+(\alpha) = \begin{cases} f(\alpha) & \text{for} \quad \alpha \geq 0 \\ 0 & \text{for} \quad \alpha < 0 \end{cases} , \quad f_-(\alpha) = \begin{cases} 0 & \text{for} \quad \alpha \geq 0 \\ f(\alpha) & \text{for} \quad \alpha < 0 \end{cases} ,$$

together with the Harten–Hyman–Lax choice of $\tilde{S}$, is an example. (6.24) then describes the operators $\hat{A}_n$ $(n = 1, 2, \ldots)$ in (6.17),[111] and hence also the operators $A_n(\Phi)$ via (6.18). The method simulates an important property of weak solutions of scalar problems of type (2.1) provided the CFL condition is fulfilled, namely their *monotonicity,* and hence their *total variation diminishing (TVD)* nature and thus their *total variation boundedness (TVB).*

We interrupt the proof of our convergence theorem for some explanations.

A method used for the numerical treatment of scalar conservation laws of type (2.1), represented in the general form

$$v^{n+1}_j = H\left(v^n_{j-k}, v^n_{j-k+1}, \ldots, v^n_{j+k}\right) ,$$

is said to be *monotone* if the partial derivatives of the function H with respect to all of its variables are nonnegative.

This imitates the following property of the weak solutions of (2.1). If two initial functions $\tilde{v}_0$ and v_0 fulfill the inequality

$$\tilde{v}_0(x) \geq v_0(x) \quad \forall x \in \mathbb{R} ,$$

this inequality holds everywhere in Ω for the weak solutions $\tilde{v}(x, t)$ and $v(x, t)$ belonging to these initial functions, respectively.

108) See footnote 97

109) It follows from this explicitness that $\Theta_n \neq \emptyset$ $(n = 1, 2, \ldots)$.

110) B. Engquist, S. Osher: Math. Comput. **34** (1980) 45–75

111) $\hat{a}_n = 0$ $(n = 1, 2, \ldots)$

In this context (and formulated for systems as well as scalar equations), *TV-boundedness* means that $\sum_j \|V^n_{j+1} - V^n_j\|_{\mathbb{R}^m} \leq K$ for all n, and *TVD methods* introduced by Harten[112] simulate the fact that the total variation of weak solutions of scalar problems does not increase, and it is obvious that TVD implies TVB. The fact that monotonicity of a scheme implies that it has TVD properties was shown in an appendix to the paper of Harten, Hyman and Lax (cf. footnote 107), written by Barbara Keyfitz.

We now resume our proof of the convergence theorem.

LeVeque[113] showed that Θ_n is discretely compact[114] provided that v_0 shows bounded variation. It is here that we put our requirement $v_0 \in BV(\mathbb{R})$ to use.

Thus, assumption (ii) in Theorem 6.1 is fulfilled because (ii*) is fulfilled.

Property (i), i.e.,

$$[\{A_n(\Phi)\}, A(\Phi)] \text{ discretely closed },$$

remains to be shown in order to demonstrate the convergence of the method using Theorem 6.1. However, in the case of TVB methods for scalar equations, this can immediately be deduced from the *Lax–Wendroff theorem.*[115]

This ends the proof for the existence of an entropy solution to our scalar conservation law example.

As a particular result, we note that a convergent monotonic consistent scheme in conservation form converges to the entropy solution of the original problem.

The Lax–Wendroff theorem referred to in the proof of the convergence theorem reads as follows.

Theorem 6.4

Let $\lambda = \frac{\Delta t}{\Delta x} = \text{const}$ and let $\Delta t = \mathcal{O}\left(\frac{1}{n}\right)$ $(n \in \mathbb{N})$ for $n \to \infty$. Define $V^n \in L_1^{\text{loc}}$ to be the step function

$$V^n(x,t) := V^n_j \;\; \textit{for} \;\; x_j - \frac{\Delta x}{2} \leq x < x_j + \frac{\Delta x}{2}\,, \qquad t_n \leq t < t_{n+1}\,,$$

$$(j = 0, \pm 1, \pm 2, \ldots; \quad n = 0, 1, 2, \ldots) \;\; \textit{on} \;\; \Omega = \{(x,t) \mid x \in \mathbb{R}\,,\; t \geq 0\}\,,$$

where the numerical values V^n_j were computed via an explicit TV-bounding finite difference one-step scheme whose numerical flux $\mathbf{g}$ is consistent and Lipschitz-continuous, and which starts from

$$V^0_j := \frac{1}{\Delta x} \int\limits_{x_j - \frac{\Delta x}{2}}^{x_j + \frac{\Delta x}{2}} V_0(\xi)\, d\xi \quad (j = 0, \pm 1, \pm 2, \ldots)\,.$$

112) A. Harten: SIAM J. Numer. Anal. **21** (1984) 1–23

113) R.J. LeVeque: *Numerical methods for conservation laws.* Basel: Birkhäuser 1990, p. 164

114) However, LeVeque used another terminology.

115) P.D. Lax, B. Wendroff: Comm. Pure Appl. Math. **13** (1960) 217–237

Assume that $\boldsymbol{V}^n$ converges for $n \to \infty$, i.e., for $\Delta t \to 0$, to a function $\boldsymbol{V}$ in the L_1^{loc}-topology, i.e., with respect to the L_1-norm on each compact subset of Ω and for each of the components of $\boldsymbol{V}^n$.

Then $\boldsymbol{V}$ is a weak solution of (2.1).

We omit the proof of the Lax–Wendroff theorem, but note that this is not a convergence theorem yet! The theorem only says that the limits of convergent sequences of approximate solutions – *assuming they exist* – are weak solutions of the original problem.

In order to complete a proof that demonstrates the important property that a method converges, one must look for properties which ensure the validity of the assumptions of Theorems 6.1 and 6.2 (as was done in the case of the Engquist–Osher scheme).

It should be mentioned that obviously the general convergence theorem of this chapter applies not only to initial value problems for scalar conservation laws, but also to certain nonlinear elliptic boundary value problems that are numerically treated using projection methods, etc.[116)]

116) R. Ansorge, J. Lei: ZAMM **71** (1991) 207–221

7
Types of Discretization Principles

7.1
Some General Remarks

In this chapter we are going to describe the principal ways of discretizing hyperbolic conservation laws, as well as their common features and differences between them.[117] For the sake of simplicity, we consider the generic scalar equation

$$L(u) := \partial_t u + \partial_x u + \partial_y u = 0 . \tag{7.1}$$

There are three basic classes of discretization method for (7.1), namely the *finite element method* (FEM), the *finite volume method* (FVM), and the *finite difference method* (FDM), already introduced briefly in Section 4.1 and used for theoretical purposes in Section 6.3. Although these names sound quite definitive, there are great many particular schemes in each of the three method classes. In the FDM, we begin by discretizing time

$$\mathbb{G}^n := \{n\Delta t \mid \Delta t > 0 , \ n \in \mathbb{N} \cup \{0\}\}$$

as well as space

$$\mathbb{G}_{i,j} := \{(i\Delta x, j\Delta y) \mid \Delta x, \Delta y > 0 , \ i,j \in \mathbb{Z}\}$$

and replace the differential operators in (7.1) by *difference operators*, for example:

$$\begin{aligned}\partial_t u(i\Delta x, j\Delta y, n\Delta t) &\approx \frac{u(i\Delta x, j\Delta y, (n+1)\Delta t) - u(i\Delta x, j\Delta y, n\Delta t)}{\Delta t} \\ &=: \frac{u_{i,j}^{n+1} - u_{i,j}^n}{\Delta t}\end{aligned}$$

and

$$\begin{aligned}\partial_x u(i\Delta x, j\Delta y, n\Delta t) &\approx \frac{u(i\Delta x, j\Delta y, n\Delta t) - u((i-1)\Delta x, j\Delta y, n\Delta t)}{\Delta x} \\ &=: \frac{u_{i,j}^n - u_{i-1,j}^n}{\Delta x}\end{aligned}$$

117) See also Chapter 6

Mathematical Models of Fluid Dynamics. R. Ansorge and T. Sonar

ISBN: 978-3-527-40774-3

$$\partial_y u(i\Delta x, j\Delta y, n\Delta t) \approx \frac{u(i\Delta x, j\Delta y, n\Delta t) - u(i\Delta x, (j-1)\Delta y, n\Delta t)}{\Delta y}$$
$$=: \frac{u^n_{i,j} - u^n_{i,j-1}}{\Delta y} .$$

Replacing the derivatives in (7.1) by the differences leads to the difference equation

$$L^h(u) := \frac{u^{n+1}_{i,j} - u^n_{i,j}}{\Delta t} + \frac{u^n_{i,j} - u^n_{i-1,j}}{\Delta x} + \frac{u^n_{i,j} - u^n_{i,j-1}}{\Delta y} \approx 0 ,$$

where the superscript h indicates a discrete value (it is very common to encounter $\Delta x = \Delta y$ and simply call this value h). Since we decided to approximate the spatial derivatives at time step $n\Delta t$, the resulting difference equation is *explicit*, i.e., the solution at the resulting time step $(n+1)\Delta t$ is explicitly computable through

$$u^{n+1}_{i,j} \approx u^n_{i,j} - \frac{\Delta t}{\Delta x}\left(u^n_{i,j} - u^n_{i-1,j}\right) - \frac{\Delta t}{\Delta y}\left(u^n_{i,j} - u^n_{i,j-1}\right) . \tag{7.2}$$

If we had decided to take the spatial differences at $t = (n+1)\Delta t$, we would have ended up with an *implicit* equation, requiring the solution of a system of linear equations at each time step. Note that in (7.2) we still used the exact solution u of (7.1). Since there will be errors due to the replacement of differential quotients by difference quotients, we cannot assume that (7.2) will hold for u with equals sign. Thus, (7.2) should be rewritten as

$$U^{n+1}_{i,j} = U^n_{i,j} - \frac{\Delta t}{\Delta x}\left(U^n_{i,j} - U^n_{i-1,j}\right) - \frac{\Delta t}{\Delta y}\left(U^n_{i,j} - U^n_{i,j-1}\right) , \tag{7.3}$$

and (7.3) now defines an approximation of $U^n_{i,j}$ to $u^n_{i,j}$ within the bounds of the discretization and other errors. Also note that the numerical solution is represented *pointwise* within the FDM framework, i.e., we know pointwise values of approximations of the exact solution.

We conclude with a philosophical remark:

> In FDMs, one *discretizes the operators* that occur in the differential equation. The discrete solution $U^h := (U^n_{i,j} | i, j \in \mathbb{Z}\ n \in \mathbb{N})$ consists of point values at the grid points of $\mathbb{G}^n \times \mathbb{G}_{i,j}$.

In the FEM, one always starts from a weak (variational) formulation like (2.2). The space $\Omega := \mathbb{R}^2 \times \{t \in \mathbb{R} \mid t \geq 0\}$ is divided into certain *elements* using a grid. The chosen elements may be the space-time *bricks* used to build the finite difference grid $\mathbb{G}^n \times \mathbb{G}_{i,j}$, but they can also be far more general. Due to their simplicity and their ability to resolve even complex geometries, triangulations are often used when the elements are triangles or tetrahedra. Applying the notion of weak solutions and choosing a space W of test functions leads to

$$\forall \Phi \in W: \quad \int_\Omega \left\{u\partial_t\Phi + u\partial_x\Phi + u\partial_y\Phi\right\} \,\mathrm{d}t\,\mathrm{d}x\,\mathrm{d}y = 0 .$$

If the test function space is replaced by a finite-dimensional subspace W^h, the relation

$$\forall \Phi^h \in W^h : \quad \int_\Omega \left\{ u\partial_t \Phi^h + u\partial_x \Phi^h + u\partial_y \Phi^h \right\} \, \mathrm{d}t \, \mathrm{d}x \, \mathrm{d}y = 0 \, .$$

must be fulfilled.

This is a set of as many equations as the dimension of $W^h := span\{\Phi_1, \ldots, \Phi_n\}$.

A popular choice is the space of polynomials linear in t and with x, y vanishing outside each element $\Sigma \subset \Omega$.

The choice of

$$U(x, y, t) := \sum_{i=1}^{n} U_i \Phi_i(x, y, t)$$

as an approximation for u then leads to a linear system for the unknown coefficients U_i since

$$\begin{aligned}
&\int_\Omega \{U\partial_t \Phi_j + U\partial_x \Phi_j + U\partial_y \Phi_j\} \, \mathrm{d}t \, \mathrm{d}x \\
&= \int_\Omega \left\{ \sum_{i=1}^{n} U_i \Phi_i \partial_t \Phi_j + \sum_{i=1}^{n} U_i \Phi_i \partial_x \Phi_j + \sum_{i=1}^{n} U_i \Phi_i \partial_y \Phi_j \right\} \mathrm{d}t \, \mathrm{d}x \, \mathrm{d}y \\
&= \sum_{i=1}^{n} U_i \left(\int_\Omega \Phi_i \partial_t \Phi_j \, \mathrm{d}t \, \mathrm{d}x \, \mathrm{d}y \right) + \sum_{i=1}^{n} U_i \left(\int_\Omega \Phi_i \partial_x \Phi_j \, \mathrm{d}t \, \mathrm{d}x \, \mathrm{d}y \right) \\
&\quad + \sum_{i=1}^{n} U_i \left(\int_\Omega \Phi_i \partial_y \Phi_j \, \mathrm{d}t \, \mathrm{d}x \, \mathrm{d}y \right) \\
&= \sum_{i=1}^{n} \left(\int_\Omega \Phi_i (\partial_t \Phi_j + \partial_x \Phi_j + \partial_y \Phi_j) \, \mathrm{d}t \, \mathrm{d}x \, \mathrm{d}y \right) U_i = 0 \, , \\
&\qquad (\forall j = 1, \ldots, n) \, .
\end{aligned}$$

FEMs of the type just described are usually called space-time FEMs. It is often convenient to only discretize in space with finite elements and tackle the time derivative in a finite difference manner. In order ro remove the coupling of every element with any other element, one can use *discontinuous Galerkin methods*, where discontinuities are allowed at element faces. This type of method is very similar to finite volume methods, but not identical. Also popular are streamline diffusion FEMs, which belong to the class of *Petrov–Galerkin* schemes. Petrov–Galerkin methods are characterized by the selection of a test space that is different from the Ansatz space in which the discrete solution is sought. Although some mathematical analysis is available for streamline diffusion methods, the numerical results obtained so far cannot cope with the results obtained by FDMs.

Again we end up with a philosophical remark:

> In FEMs, one *discretizes the solution* of the differential equation. The discrete solution consists of polynomials defined piecewise on the elements.

The third class of methods, FVMs, share all of the good properties of FDMs as well as of those FEMs. As with FEMs, arbitrary cells in space can be used for the discretization, while all of the FD technology is also applicable at the same time. Here, one starts with a type of weak solution which stems directly from gas dynamics, where the conservative equations are formulated as integral balances on so-called control volumes. For our generic model, this weak form results from integrating

$$\partial_t u + \partial_x u + \partial_y u = \partial_t u + \langle \nabla, \boldsymbol{u} \rangle = 0 \,, \quad \boldsymbol{u} := (u, u)^T$$

over a bounded control volume $\sigma \subset \mathbb{R}^2$ with boundary $\partial\sigma$, where the unit outer normal vector $\boldsymbol{n}$ is defined almost everywhere. From Gauss' divergence theorem it then follows

$$\frac{\mathrm{d}}{\mathrm{d}t} \int_\sigma u \; \mathrm{d}x \; \mathrm{d}y + \oint_{\partial\sigma} \langle \boldsymbol{u}, \boldsymbol{n} \rangle \; \mathrm{d}s = 0 \,.$$

Now, the *cell average* of u on σ is defined as

$$\overline{u}_\sigma(t) := \frac{1}{|\sigma|} \int_\sigma u(x, y, t) \; \mathrm{d}x \; \mathrm{d}y \,,$$

where $|\sigma|$ denotes the area of σ, so that the weak formulation of gas dynamics unfolds itself as an evolution equation for cell averages

$$\frac{\mathrm{d}}{\mathrm{d}t} \overline{u}_\sigma = -\frac{1}{|\sigma|} \oint_{\partial\sigma} \langle \boldsymbol{u}, \boldsymbol{n} \rangle \; \mathrm{d}s \,. \tag{7.4}$$

FVMs are derived from (7.4) by approximating the line integrals with a suitable Gauss formula. The product $\langle \boldsymbol{u}, \boldsymbol{n} \rangle$ is replaced by a so-called *numerical flux function* which is essentially a finite difference formula.

Again we conclude with a philosophical remark:

> FVMs are discrete *evolution equations for cell averages*. There is freedom in relation to representing the numerical solution.

Some people like to view FVMs as being part of the Petrov–Galerkin family of FEMs with piecewise constant trial functions. The central idea in FVMs is to *adapt FDMs to unstructured grids* like triangulations and to use the computational advantages from FDMs and FEMs.

There are even more discretization methods available than the three classes described here, but they are often classified as particular instances of FDMs, FEMs and FVMs. A method that is popularly used in the computation of incompressible viscous flows, and is now becoming more and more attractive for compressible fluid flow, is the class of *spectral discretizations*. Here the Ansatz space consists of globally defined trigonometric or polynomial functions on the whole domain (or, as in the spectral element method, on macroelements). Choosing an FEM-like Ansatz leads to *spectral Galerkin* methods. If the differential equation must be satisfied pointwise at so-called collocation points, one talks about *spectral collocation* or *pseudospectral* methods. These methods are very close to being FDMs with large stencils.

7.2 Finite Difference Calculus

Let us now study the structures of finite difference methods more intensively.

A century before the invention of calculus by Leibniz and Newton, the foundations had been laid for a discrete calculus employing differences in functions. Napier's[118] and Briggs'[119] invention of logarithms revolutionized naval navigation and necessitated interpolation formulae for computing intermediate values for tabulated data. It was the genius Thomas Harriot[120] who founded the field of finite difference calculus. This field flourished in the days of Leibniz and Euler and came to exert enormous influence in the area of the numerical treatment of ordinary and partial differential equations.

To again give the reader an idea of the spirit of finite difference calculus, we consider the *forward difference operator* Δ defined by

$$\Delta u(x) := u(x + \Delta x) - u(x) ,$$

where Δx denotes an increment that is different from zero. It is often convenient to introduce the *shift operator* S through

$$Su(x) := u(x + \Delta x)$$

so that $\Delta = S - I$, where I is the identity operator. Assuming a smooth function $x \to u(x)$, Taylor series expansion results in

$$u(x + \Delta x) = \sum_{\nu=0}^{\infty} \frac{1}{\nu!} \Delta x^{\nu} D^{\nu} u(x) ,$$

where D^{ν} is defined as $\underbrace{\frac{\mathrm{d}}{\mathrm{d}x} \circ \frac{\mathrm{d}}{\mathrm{d}x} \circ \ldots \circ \frac{\mathrm{d}}{\mathrm{d}x}}_{\nu \text{ times}}$ and $D^0 = I$.

Utilizing the shift operator, this is simply the representation

$$Su(x) = \left(\sum_{\nu=0}^{\infty} \frac{1}{\nu!} \Delta x^{\nu} D^{\nu} \right) u(x) ,$$

118) John Napier (1559–1617); Merchiston
119) Henry Briggs (1556–1630); London, Oxford
120) Thomas Harriot (1560–1621); Oxford

where now, formally, the term in parentheses is exactly $e^{\Delta x D}$, i.e., we have found the representation

$$S = e^{\Delta x D} \tag{7.5}$$

and hence

$$\Delta = e^{\Delta x D} - I \,. \tag{7.6}$$

Introducing the *central difference operator* δ by means of

$$\delta u(x) := u(x + \Delta x/2) - u(x - \Delta x/2) \,,$$

a further Taylor series

$$u(x \pm \Delta x/2) = \sum_{\nu=0}^{\infty} \frac{1}{\nu!} (\pm \Delta x)^{\nu} \frac{1}{2^{\nu}} D^{\nu} u(x)$$

reveals

$$\delta u(x) = 2 \sum_{\nu=0}^{\infty} \frac{1}{(2\nu + 1)!} \left(\frac{\Delta x}{2} D \right)^{2\nu+1} u(x) \,,$$

i.e.,

$$\delta = 2 \sinh \left(\frac{\Delta x}{2} D \right) \,. \tag{7.7}$$

Inverting this equation gives a relation between the differential operator D and the central difference operator δ, namely

$$D = \frac{2}{\Delta x} \sinh^{-1} \frac{\delta}{2} = \frac{1}{\Delta x} \left(\delta - \frac{1^2}{2^2 3!} \delta^3 + \frac{1^2 3^2}{2^4 5!} \delta^5 - \ldots \right) \tag{7.8}$$

where $\delta^n := \underbrace{\delta \circ \delta \circ \ldots \circ \delta}_{n \text{ times}}$. Equation (7.8) is an *exact difference formula* for the operator D.

Other equations like (7.8) and (7.6) can also be constructed.

We are mainly interested in the applicability of formulae like (7.8). To get an idea, we consider the simple linear transport equation $\partial_t u + \partial_x u = 0$, which we rewrite as

$$\partial_t u = L(D) u := -\partial_x u \,,$$

where the differential operator D is now a partial derivation with respect to x. From (7.5), we deduce

$$u(x, t + \Delta t) = \left(e^{\Delta t \partial_t} \right) u(x, t)$$

and, since $\partial_t = L(D)$, we conclude

$$u_i^{n+1} = \left(e^{\Delta t L(D)} \right) u_i^n \,. \tag{7.9}$$

We now use (7.8) to eliminate D in terms of δ in (7.9). This gives

$$u_i^{n+1} = \left(e^{\Delta t L(2/\Delta x \sinh^{-1}(\delta/2))}\right) u_i^n . \tag{7.10}$$

Note that (7.10) is not an approximate difference equation but an exact one. Retaining only the linear terms in the exponential function gives

$$U_i^{n+1} = \left(I + \Delta t L(2/\Delta x \sinh^{-1}(\delta/2))\right) U_i^n .$$

Retaining only linear terms in the expansion (7.8) and noting that $L(D) = -D = -\partial_x$ finally leads to

$$U_i^{n+1} = \left(I - \Delta t \frac{\delta}{\Delta x}\right) U_i^n .$$

Expanding the central difference operator gives the finite difference method

$$U_i^{n+1} = U_i^n - \frac{\Delta t}{\Delta x}\left(U_{i+1/2}^n - U_{i-1/2}^n\right) \tag{7.11}$$

for the transport equation.

Hence, finite difference calculus can be used directly to construct finite difference methods for partial differential equations. More constructions of this type can be found in the literature.

Now let the above difference scheme act on the true solution u. Again employing Taylor series expansions,

$$\begin{aligned}
u_i^{n+1} &= u(x,t) + \Delta t \partial_t u(x,t) + \mathcal{O}(\Delta t^2) \\
u_{i+1/2}^n &= u(x,t) + \frac{\Delta x}{2}\partial_x u(x,t) + \frac{1}{2!}\frac{\Delta x^2}{2^2}\partial_x^2 u(x,t) + \mathcal{O}(\Delta x^3) \\
u_{i-1/2}^n &= u(x,t) - \frac{\Delta x}{2}\partial_x u(x,t) + \frac{1}{2!}\frac{\Delta x^2}{2^2}\partial_x^2 u(x,t) + \mathcal{O}(\Delta x^3) ,
\end{aligned}$$

and inserting them into (7.11) gives

$$u + \Delta t \partial_t u + \mathcal{O}(\Delta t^2) = u - \frac{\Delta t}{\Delta x}\left(\Delta x \partial_x u + \mathcal{O}(\Delta x^3)\right)$$

which, after rearranging, leads to

$$\partial_t u + \partial_x u = \mathcal{O}(\Delta t, \Delta x^2) . \tag{7.12}$$

Thus, our simple difference method (7.11) is of *order one in time* and of *order two in space*. This notion of order is usually called the *truncation error* since it results from truncating a Taylor series.

Note that we have not said a word about the convergence of (7.11). The property that (7.12) formally tends to $\partial_t u + \partial_x u = 0$ for $\Delta t, \Delta x \to 0$ is called *consistency*.

If a difference operator is applied to a sequence $\left(U_i^n | i \in \mathbb{Z}\right)$, then it acts on the sequence index, i.e., $\Delta_i U_i^n := \Delta U_i^n = U_{i+1}^n - U_i^n$.

For later reference, we now introduce the *antidifference operator* Δ^{-1} as follows. If $\Delta u_i = g_i$ holds, then we define

$$u_i = \Delta^{-1} g_i + c ,$$

where c denotes an arbitrary constant. The constant appears because after the application of Δ we get $\Delta u_i = g_i$ back. We first prove

$$\sum_{i=n_0}^{n-1} \Delta u_i = u_n - u_{n_0} , \tag{7.13}$$

$$\Delta \left(\sum_{i=n_0}^{n-1} u_i \right) = u_n . \tag{7.14}$$

Equation (7.13) simply follows from

$$\sum_{i=n_0}^{n-1} \Delta u_i = (u_{n_0+1} - u_{n_0}) + (u_{n_0+2} - u_{n_0+1}) + \ldots + (u_{n-1} - u_{n-2}) + (u_n - u_{n-1}) ,$$

while (7.14) is simply

$$\Delta \left(\sum_{i=n_0}^{n-1} u_i \right) = \sum_{i+1=n_0}^{n-1} u_{i+1} - \sum_{i=n_0}^{n-1} u_i = \sum_{i=n_0-1}^{n-1} u_{i+1} - \sum_{i=n_0}^{n-1} u_i .$$

Using formula (7.14), we can immediately deduce

$$\Delta^{-1} u_n = \sum_{i=n_0}^{n-1} u_i + c , \tag{7.15}$$

where n_0 is an arbitrary start index.

We now provide a discrete analog of the product rule, namely

$$\Delta(u_i v_i) = S u_i \Delta v_i + v_i \Delta u_i . \tag{7.16}$$

The proof consists of evaluating the right and left hand sides to see that they coincide.

Writing the product rule in the form $v_i \Delta u_i = \Delta(u_i v_i) - u_{i+1} \Delta v_i$ and applying the operator Δ^{-1} results in $\Delta^{-1}(v_i \Delta u_i) = \Delta^{-1}\Delta(u_i v_i) - \Delta^{-1}(u_{i+1}\Delta v_i)$. Following formula (7.15), we finally get the formula for *summation by parts*, which is a discrete analog to the formula of integration by parts, namely

$$\sum_{i=n_0}^{n-1} v_i \Delta u_i = u_n v_n - \sum_{i=n_0}^{n-1} u_{i+1} \Delta v_i + c . \tag{7.17}$$

7.3 The CFL Condition

In 1928, Courant,[121] Friedrichs,[122] and Hans Lewy published a milestone paper (cf. Section 6.1) which was ignored during the war and only became famous when it was rediscovered during the Manhattan Project.

As already emphasized in the last chapter, Courant's interests at that time mainly involved proving the existence of solutions to partial differential equations. In that field he was one of the founders of the idea of employing finite difference schemes as approximations to partial differential equations. Proving the existence of solutions to partial differential equations involved proving that the difference scheme was consistent and convergent, and furthermore that the limit solution (as Δt and Δx tend to zero) satisfied the original differential equation. Although Courant, Friedrichs and Lewy succeeded with this scheme for the (elliptic) Laplace equation, they encountered serious problems when applying it to the (hyperbolic) wave equation. They finally succeeded in formulating the celebrated CFL condition, which is necessary for convergence.

In order to demonstrate the basic behavior observed in the context considered here, let us consider the simple forward explicit difference scheme for the slightly more general transport equation $\partial_t u + a\partial_x u = 0$, $a < 0$, namely

$$U_i^{n+1} = U_i^n - a\frac{\Delta t}{\Delta x}\left(U_{i+1}^n - U_i^n\right) = \left(1 + a\frac{\Delta t}{\Delta x}\right)U_i^n - a\frac{\Delta t}{\Delta x}U_{i+1}^n\,. \tag{7.18}$$

(7.18) is an explicit three-point scheme for the explicit computation of U_i^{n+1}. Viewed in the (x, t)-plane, we can identify the *difference stencil* of (7.18), as shown in Fig. 7.1.

The stencil gives rise to the definition of *numerical characteristics*, which are straight lines enveloping the stencil. The numerical characteristics cut off an interval of the axis $t = 0$, which is called the *numerical domain of dependence*. Changing the initial datum in this interval changes the numerical solution at point P, while perturbations outside this interval would not affect the numerical solution at P. However, there is also the characteristic Γ of the differential equation that we would like to approximate. If Γ lies outside the numerical characteristics, as shown in Fig. 5.8, then changing the initial datum between Γ and the numerical characteristic would change the true solution at P but not the numerical solution. Thus, convergence of the difference scheme is not possible. Keeping Γ inside the numerical characteristics ensures that all perturbations of the initial datum which affect the true solution at P will always affect the numerical solution at P too.

The characteristic Γ has the slope

$$\frac{\mathrm{d}t}{\mathrm{d}x} = \frac{1}{a}$$

121) Richard Courant (1888–1972); Göttingen, New York

122) Kurt Otto Friedrichs (1901–1982); Brunswick, New York

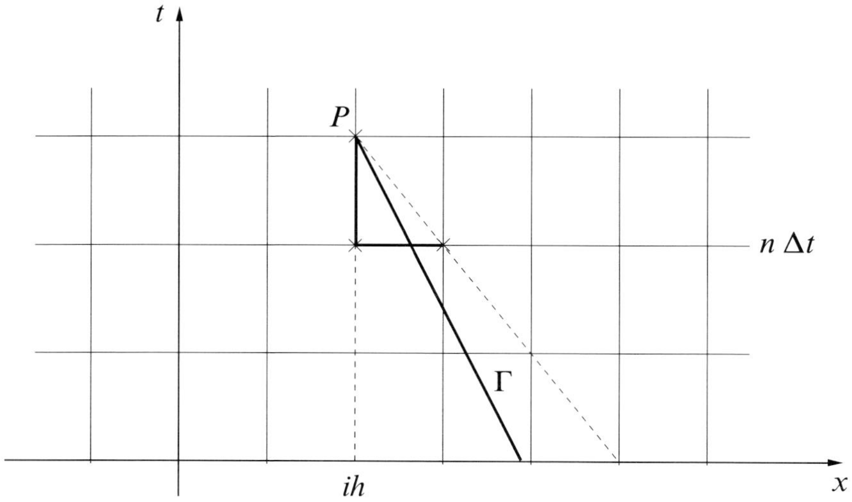

Fig. 7.1 Difference stencil and characteristics.

and the right numerical characteristic has a slope of $-\Delta t/\Delta x$. Thus, $\Delta t/\Delta x \le 1/|a|$ is necessary for convergence, which is already the Courant–Friedrichs–Lewy (CFL) condition

$$\Delta t \le \frac{\Delta x}{|a|} . \tag{7.19}$$

In the case of $a > 0$, and replacing the forward difference equation by the backward one, (7.19) again ensures appropriate behavior for the domains of dependence. This dependence of the type of difference equation on the sign of a, i.e., on the direction of movement of the wave, leads to so-called *upwind schemes*.

In the case of a scalar conservation law $\partial_t u + \partial_x f(u) = 0$, the same arguments hold for the quasilinear form $\partial_t u + f'(u)\partial_x u = 0$. Hence, instead of (7.19), explicit finite difference schemes with a three-point stencil have to satisfy the *CFL condition*

$$\frac{\Delta t}{\Delta x} \max_i |f'(U_i^n)| \le 1 \tag{7.20}$$

at each time step n.

7.4 Lax–Richtmyer Theory

We now formalize the treatment of finite difference schemes, supplementing Chapter 6, in order to present at least the basic historical idea of a convergence theory for this particular type of scheme.

Although we have to confine ourselves to linear problems, the Lax–Richtmyer theory allows us to gain deep insights into the interplay between the notions of consistency, stability and convergence. It was originally formulated for autonomous, linear initial value problems, but was later generalized by several authors to include nonautonomous and nonlinear problems.[123)]

We consider an *abstract evolution equation* in a Banach space B, where a linear operator $A\colon D_A \longrightarrow B$ acts upon a subspace D_A of B. The evolution equation we want to discretize is then given as

$$\begin{aligned} \frac{\mathrm{d}}{\mathrm{d}t} u(t) &= A(u(t)), \quad t \in [0, T]\,, \ T > 0 \\ u(0) &= u_0, \quad u_0 \in D_A\,. \end{aligned} \tag{7.21}$$

A one-parameter family $u(t)$ with $t \in [0, T]\,,\ u(t) \in D_A$, is called a *genuine solution* of (7.21), if

$$\lim_{\substack{\Delta t \to 0 \\ 0 \le t \le T}} \left\| \frac{u(t + \Delta t) - u(t)}{\Delta t} - Au(t) \right\|_B = 0\,. \tag{7.22}$$

Let D be the set of all initial data $u_0 \in B$ so that a unique genuine solution exists with $u(0) = u_0$ and such that convergence in (7.22) is uniform in t.[124)] There is then a one-parameter family of linear operators $E_0(t)\colon D \longrightarrow B$ defined by

$$u(t) = E_0(t) u_0$$

which are called *evolution operators* or *solution operators*. We always assume that our problem is properly posed, which in particular implies that $E_0(t)$ is bounded uniformly in t, i.e., there is a constant $K > 0$ such that $\|E_0(t)\|_{D \to B} < K$ for $0 \le t \le T$. Then the Hahn–Banach extension theorem implies the existence of a *generalized solution operator* $E(t)\colon B \longrightarrow B$ with the same bound K as $E_0(t)$. The generalized solution operator has the *semigroup property*

$$\forall\, s, t \ge 0: \quad E(s + t) = E(s)\, E(t)$$

and the closure of the operator A, which we also denote by A, is the *infinitesimal generator* of this semigroup. One can further show that $E(t)$ commutes with A, so that

$$E(t) A u_0 = A E(t) u_0\,.$$

Then (7.21) can be written as

$$\lim_{\substack{\Delta t \to 0 \\ 0 \le t \le T}} \left\| E(t) \left(\frac{E(\Delta t) - I}{\Delta t} - A \right) u_0 \right\|_B = 0\,,$$

which, since $\|E(t)\|_B < K$ for $0 \le t \le T$, implies uniform convergence of (7.22) with respect to t.

123) See R. Ansorge: Survey of equivalence theorems in the theory of difference approximations for partial initial value problems. In: J.J. Miller (ed.): *Topics in numerical analysis.* London: Academic Press 1977

124) We shall see shortly that uniform convergence is automatic.

In order to discuss fairly general finite difference schemes, let us introduce linear *difference operators*

$$\Lambda_{0/1} = \Lambda_{0/1}(\Delta t, \Delta x_1, \ldots, \Delta x_d)$$

and the class of difference schemes

$$\Lambda_1 U^{n+1} = \Lambda_0 U^n .$$

We assume the existence of Λ_1^{-1} and that $\Lambda_1^{-1}\Lambda_0$ is a bounded linear operator defined on B. From the CFL condition, we know that Δt and Δx_i cannot go to zero independently of each other as far as explicit methods are concerned. Hence, we assume the relation $\Delta x_i = \varphi_i(\Delta t), i = 1, \ldots, d$ and introduce the *discrete evolution operator*

$$C(\Delta t) := \Lambda_1^{-1}\left(\Delta t, \varphi_1(\Delta t), \ldots, \varphi_d(\Delta t)\right) \cdot \Lambda_0\left(\Delta t, \varphi_1(\Delta t), \ldots, \varphi_d(\Delta t)\right)$$

so that the class of difference schemes under consideration is

$$U^{n+1} = C(\Delta t)\, U^n . \tag{7.23}$$

The family of operators $C(\Delta t)$ is said to define a *consistent* approximation of the initial value problem if, for every $u(t)$ in a class of genuine solutions arising from initial data $u(0)$ out of a set that is dense in B, the *consistency condition*

$$\lim_{\substack{\Delta t \to 0 \\ 0 \le t \le T}} \left\| \left(\frac{C(\Delta t) - I}{\Delta t} - A \right) u(t) \right\|_B = 0 \tag{7.24}$$

holds. Since $u(t)$ denotes a genuine solution, we can use the fact that, due to (7.22), $Au(t)$ is close to $(u(t+\Delta t) - u(t))/\Delta t$ in order to get the alternative definition

$$\lim_{\substack{\Delta t \to 0 \\ 0 \le t \le T}} \left\| \frac{u(t+\Delta t) - C(\Delta t)u(t)}{\Delta t} \right\|_B = 0 . \tag{7.25}$$

The quantity

$$\sup_{t \in [0,T]} \left\| \frac{u(t+\Delta t) - C(\Delta t)u(t)}{\Delta t} \right\|_B := \hat{\varepsilon}(\Delta t, u_0)$$

is called *truncation error*, and if this error is of order p, i.e.,

$$\hat{\varepsilon}(\Delta t, u_0) = \mathcal{O}(\Delta t^p) ,$$

the method is called a *method of order p*. Consistency therefore means

$$\hat{\varepsilon}(\Delta t, u_0) = \mathrm{o}(1)$$

which is fulfilled in particular for $p > 0$.

Note that this abstract notion is in complete agreement with our earlier calculations in (7.12).

Consistency guarantees that, in the limit $\Delta t \to 0$, the difference scheme formally reproduces the differential equation. The notion of convergence is concerned with the limit behavior of the discrete solutions. *Consistency does not necessarily already imply convergence!*

Due to (7.23), the iterated application of $C(\Delta t)$ gives

$$U^n = C(\Delta t)\, U^{n-1} = C(\Delta t) \circ C(\Delta t)\, U^{n-2} = \ldots = C^n(\Delta t)\, U^0 = C^n(\Delta t) u_0 \,,$$

and this discrete function is expected to be a good approximation to $u(n\Delta t) = E(n\Delta t)u_0$. The family of operators $C(\Delta t)$ provides a *convergent* approximation to the initial value problem if, for fixed $t \in [0, T]$,

$$\lim_{j\to\infty} \left\| C(\Delta_j t)^{n_j} u_0 - E(t) u_0 \right\|_B = 0 \tag{7.26}$$

holds for each sequence $(\Delta_j t)_{j=1,2,\ldots}$ tending to zero, where n_j is chosen to satisfy $\lim_{j\to\infty} n_j \Delta_j t = t$.

In reaching the discrete solution at time $t = n\Delta t$, the operator $C(\Delta t)$ has to be applied n times on the initial datum. If we consider small errors in u_0 or errors due to rounding during the computation, we would not be satisfied with a difference scheme which would amplify these small perturbations until they spoil the solution. The central idea of numerical stability is therefore to put bounds on the possible amplification of initial data. The operator $C(\Delta t)$ is said to define a *stable* approximation if, for some $t^* > 0$, the infinite set of operators

$$C(\Delta t)^n \,, \quad 0 < \Delta t < t^* \,, \quad 0 \le n\Delta t \le T \,,$$

are uniformly bounded, i.e., if there is a constant κ with

$$\| C(\Delta t)^n \|_B \le \kappa \,, \forall\, n \in \mathbb{N} \,, \forall\, \Delta t \quad \text{with } 0 < \Delta t \le t^* \quad \text{and with} \quad 0 \le n\Delta t \le T \,.$$

With the notions of consistency, stability and convergence, we can now formulate the central *equivalence theorem* of the Lax–Richtmyer theory.

Theorem 7.1

A finite difference scheme approximating a properly posed initial value problem is convergent if and only if it is consistent and stable.

Proof (of the direction: *consistency* $\wedge$ *stability* $\Rightarrow$ *convergence*[125]):

$U(0) = u_0$ implies

$$U^n - u(t) = \sum_{\nu=1}^{n} C(\Delta t)^{\nu-1} \left\{ C(\Delta t)\, u\left(t_{n-\nu}\right) - u\left(t_{n-\nu+1}\right) \right\} + u(t_n) - u(t)$$

$$(t_\mu = \mu\, \Delta t) \,.$$

125) From an engineering point of view, this is the more important direction. The complete proof can be found in the original paper of Lax and Richtmyer cited in footnote 99. If consistency and stability are defined properly, theorems of the Lax–Richtmyer type can even be proven in the case of nonlinear difference schemes for nonlinear equations, as previously mentioned in footnote 123.

Thus, from the stability property, we obtain

$$\begin{aligned}\|U^n - u(t)\|_B &\le \sum_{\nu=1}^{n} \left\|C(\Delta t)^{\nu-1}\right\|_B \cdot \left\|C(\Delta t)\, u\,(t_{n-\nu}) - u\,(t_{n-\nu+1})\right\|_B \\ &\quad + \|u(t_n) - u(t)\|_B \\ &\le \kappa \sum_{\nu=1}^{n} \left\|C(\Delta t)\, E_0\,(t_{n-\nu})\; u_0 - E_0\,(t_{n-\nu+1})\; u_0\right\|_B \\ &\quad + \|u(t_n) - u(t)\|_B\ ,\end{aligned}$$

leading to

$$\|U^n - u(t)\|_B \le \kappa\, n\, \Delta t \cdot \hat{\varepsilon}(\Delta t, u_0) + \|u(t_n) - u(t)\|_B\ .$$

Taking $n\,\Delta t \le T$ into account, we find

$$\|U^n - u(t)\|_B \le \kappa\, T \hat{\varepsilon}(\Delta t, u_0) + \|u(n\,\Delta t) - u(t)\|_B\ .$$

Because of $n\,\Delta t \ \rightarrow\ t$, because of the continuity of u with respect to t, and because of the consistency condition (7.25), the truth of the assertion is obvious.

Particularly for $n\,\Delta t \ = \ t$, we get an estimate for the *global discretization error* $\|U^n - u(t)\|_B$, namely

$$\|U^n - u(t)\|_B \le \kappa\, t\, \hat{\varepsilon}(\Delta t, u_0)\ .$$

Round-off errors are not included here!

Example

Let us again treat the transport equation $\partial_t u + a\partial_x u = 0$ in the case of $a < 0$.

Thus, if (7.18) is expressed in terms of the Lax–Richtmyer theory, and if a relation between the step sizes Δt and Δx is described, we find

$$\|C(\Delta t)\|_\infty \le \left|1 + a\frac{\Delta t}{\Delta x}\right| + |a|\frac{\Delta t}{\Delta x}\ .$$

The CFL condition (7.19) then ensures that

$$\|C(\Delta t)\|_\infty \le \left(1 + a\frac{\Delta t}{\Delta x}\right) + |a|\frac{\Delta t}{\Delta x} = 1\ ,$$

and hence

$$\|C^n(\Delta t)\|_\infty \le \|C(\Delta t)\|_\infty^n \le 1\ .$$

Thus, the CFL condition guarantees the numerical stability of the scheme in the sense of the Lax–Richtmyer theory.

The other direction is also true: stability in the sense of the Lax–Richtmyer theory requires the CFL condition to be fulfilled, and this holds independently of the norm.

Of course, the upwind scheme for $a > 0$ can be treated in the same way.

The upwind idea can also be looked at from the point of view of information theory: looking in the direction of information travel, it seems unreasonable to use values for the present time step whose positions lie downstream and have not yet been influenced by the information used to compute a future value of the solution at a given position in space.

7.5 The von Neumann Stability Criterion

The consistency of a finite difference scheme is easy to verify via Taylor expansions. In contrast, it may be much harder to verify stability using the pure definition. In the case of linear initial value problems or problems with periodic boundary data, there is a simple mechanism that was derived by von Neumann.[126)]

We start with a complex Fourier transform of a discrete function represented as a sequence $U^n = (U_j^n)_{j\in\mathbb{Z}}$ and defined at grid points only. The discrete Fourier transform is defined as

$$\widehat{U^n}(\xi) = \sum_{j\in\mathbb{Z}} U_j^n e^{-\mathrm{i}j\xi}, \quad 0 \le \xi < 2\pi .$$

The importance of the discrete Fourier transform in analyzing the stability of a finite difference scheme can be seen if it is applied to the shift operator S defined in Section 7.2. For our purposes, we refine the definition by calling $S_+ U^n := S(\Delta x) U^n := (U_{j+1}^n)_{j\in\mathbb{Z}}$ the *forward shift operator*. Therefore, it follows from the definition of the discrete Fourier transform that

$$\begin{aligned}
\widehat{S_+ U^n}(\xi) &= \sum_{j\in\mathbb{Z}} U_{j+1}^n e^{-\mathrm{i}j\xi} = \sum_{j\in\mathbb{Z}} U_j^n e^{-\mathrm{i}(j-1)\xi} \\
&= \sum_{j\in\mathbb{Z}} U_j^n e^{-\mathrm{i}j} e^{\mathrm{i}\xi} = e^{\mathrm{i}\xi} \sum_{j\in\mathbb{Z}} U_j^n e^{-\mathrm{i}j\xi} \\
&= e^{\mathrm{i}\xi}\, \widehat{U^n}(\xi) .
\end{aligned}$$

Thus, applying the shift operator to a grid function is, in Fourier space, multiplication with $e^{\mathrm{i}\xi}$. The function $e^{\mathrm{i}\xi}$ is called the *symbol* of the shift operator S.

Completely analogously, one obtains the symbol $e^{-\mathrm{i}\xi}$ for the *backward shift operator* $S_- U^n := S(-\Delta x) U^n$, i.e.,

$$\widehat{S_- U^n}(\xi) = e^{-\mathrm{i}\xi}\, \widehat{U^n}(\xi) .$$

126) John von Neumann (1903–1957); Berlin, Hamburg, Göttingen, Princeton

Since every linear finite difference scheme is made up of combinations of forward and backward shift operators, we now have a way to look at discrete Fourier transforms of difference schemes.

Example

Again consider the difference scheme

$$U_i^{n+1} = U_i^n - \frac{\Delta t}{2\Delta x} a \left(U_{i+1}^n - U_{i-1}^n \right)$$

as an approximation to $\partial_t u + a\partial_x u = 0$. It is an easy exercise to see that the scheme is second-order consistent in space and first-order consistent in time. Written in terms of shift operators and sequences, we get

$$U^{n+1} = \left(I - \frac{\Delta t}{2\Delta x} a(S_+ - S_-) \right) U^n .$$

Now, taking Fourier transforms on both sides, it follows that

$$\widehat{U^{n+1}}(\xi) = \left(1 - \frac{\Delta t}{2\Delta x} a \left(e^{\mathrm{i}\xi} - e^{-\mathrm{i}\xi} \right) \right) \widehat{U^n}(\xi) .$$

If the symbol $\Theta(\Delta t, \xi) := 1 - \frac{\Delta t}{2\Delta x} a \left(e^{\mathrm{i}\xi} - e^{-\mathrm{i}\xi} \right)$ is found to be larger than modulus one, this means that any small perturbation in U^n will be amplified in the step computing U^{n+1}. In our case, since $e^{\pm \mathrm{i}\xi} = \cos\xi \pm \mathrm{i}\sin\xi$, we get

$$\Theta(\Delta t, \xi) = 1 - \frac{\Delta t}{\Delta x} a\, i\, \sin\xi .$$

Taking the modulus, this amounts to

$$|\Theta(\Delta t, \xi)|^2 = 1 + \left(\frac{\Delta t}{\Delta x} a \right)^2 \sin^2\xi \geq 1 ,$$

regardless of how small the selected time step is. This symbol is usually called the *amplification factor* of the difference scheme. The present scheme is unconditionally unstable, since the modulus of the amplification factor is larger than one except for $\xi \in \{0, \pi, 2\pi\}$. Assuming $a > 0$, the upwind difference scheme

$$U_i^{n+1} = U_i^n - \frac{\Delta t}{\Delta x} a \left(U_i^n - U_{i-1}^n \right)$$

results in the sequence formulation

$$U^{n+1} = \left(I - \frac{\Delta t}{\Delta x} a(I - S_-) \right) U^n$$

which gives the discrete Fourier transform

$$\widehat{U^{n+1}}(\xi) = \left(1 - \frac{\Delta t}{\Delta x} a(1 - e^{-\mathrm{i}\xi}) \right) \widehat{U^n}(\xi) .$$

Let us define $K := \frac{\Delta t}{\Delta x} a$ and compute the amplification factor

$$\Theta(\Delta t, \xi) = \left|1 - K(1 - e^{-\mathrm{i}\xi})\right| \leq |1 - K| + K \underbrace{|e^{-\mathrm{i}\xi}|}_{=1}$$

$$\stackrel{\text{CFL condition}}{\leq} 1 - K + K = 1 \, .$$

Hence, our upwind scheme is stable under the usual CFL condition.

We remark in passing that the same analysis holds in multiple spatial dimensions. Here, the discrete Fourier transform is defined as

$$\widehat{U^n}(\Xi) = \sum_{k \in \mathbb{Z}^d} U_k^n e^{\mathrm{i}k^T \Xi}$$

where d is the space dimension and $\Xi = (\xi_1, \ldots, \xi_d)^T$.

In the case of systems of equations with n components in d spatial dimensions, amplification is given by

$$\widehat{\boldsymbol{U}^{n+1}}(\Xi) = \boldsymbol{\Theta}(\Delta t, \Xi)\widehat{\boldsymbol{U}^n}(\Xi)$$

where $\boldsymbol{\Theta}$ is an $n \times n$ matrix, the *amplification matrix*. For reasons of generality, we stay with this case (systems of equations) in the following.

Recursion reveals the relation

$$\widehat{\boldsymbol{U}^n}(\Xi) = \boldsymbol{\Theta}^n(\Delta t, \Xi)\widehat{\boldsymbol{U}^0}(\Xi)$$

so that stability is associated with the properties of the iterated amplification matrix. In general, we can state the *necessary and sufficient stability criterion*, that for some positive Δt^* the matrices (or, in the case of scalar equations, the amplification factor)

$$\boldsymbol{\Theta}^n(\Delta t, \Xi)$$

should be uniformly bounded for $0 < \Delta t < \Delta t^*, 0 \leq n\Delta t \leq T$ and all $\Xi \in \mathbb{Z}^d$. If $\varrho_{\boldsymbol{\Theta}}(\Delta t, \Xi)$ denotes the spectral radius of $\boldsymbol{\Theta}$ (i.e., the maximum of the moduli of the eigenvalues), then it follows from linear algebra that

$$\varrho^n_{\boldsymbol{\Theta}}(\Delta t, \Xi) \leq \|\boldsymbol{\Theta}^n(\Delta t, \Xi)\| \leq \|\boldsymbol{\Theta}(\Delta t, \Xi)\|^n \, .$$

A necessary condition for stability is therefore the existence of a constant C such that

$$\varrho^n_{\boldsymbol{\Theta}}(\Delta t, \Xi) \leq C$$

for $0 < \Delta t < \Delta t^*, 0 \leq n\Delta t \leq T$ and all $\Xi \in \mathbb{Z}^d$. Without loss of generality, we assume that $C \geq 1$ and therefore

$$\varrho_{\boldsymbol{\Theta}}(\Delta t, \Xi) \leq C^{\frac{1}{n}} \, , \quad 0 \leq n \leq \frac{T}{\Delta t} \, .$$

Thus, in particular, we get

$$\varrho_{\boldsymbol{\Theta}}(\Delta t, \Xi) \leq C^{\frac{\Delta t}{T}} .$$

Now the expression $C^{\Delta t/T}$ is bounded for $0 < \Delta t < \Delta t^*$ by a linear function $1 +$ const Δt.

Thus, a necessary condition for stability is the celebrated *von Neumann condition,*

$$|\lambda_i| \leq 1 + \mathcal{O}(\Delta t) ,$$

for $0 < \Delta t < \Delta t^*$, all $\Xi \in \mathbb{Z}^d$ and $i = 1, \ldots, n$, where λ_i denotes the i-th eigenvalue of $\boldsymbol{\Theta}(\Delta t, \Xi)$.

It should be noted that there are several other stability conditions, like the famous Kreiss matrix theorem, or the much less known criterion of Buchanan.[127)]

7.6 The Modified Equation

Finally, we briefly describe a tool that can often help in the analysis of finite difference procedures.

For every finite difference approximation of order $\mathcal{O}(\Delta t^p, \Delta x^q)$ of a given differential equation, there is another differential equation to which the difference scheme is a better approximation.

In order to demonstrate this remark, let us start again from the simple upwind scheme

$$U_i^{n+1} = U_i^n - \frac{\Delta t}{\Delta x} a \left(U_i^n - U_{i-1}^n \right) \tag{7.27}$$

for the transport equation

$$\partial_t u + a \partial_x u = 0 , \quad a > 0 . \tag{7.28}$$

Inserting the Taylor series expansions for a sufficiently smooth function u, i.e.,

$$u_i^{n+1} = u + \Delta t \partial_t u + \frac{\Delta t^2}{2} \partial_t^2 u + \frac{\Delta t^3}{6} \partial_t^3 u + \mathcal{O}(\Delta t^4)$$

$$u_{i-1}^n = u - \Delta x \partial_x u + \frac{\Delta x^2}{2} \partial_x^2 u - \frac{\Delta x^3}{6} \partial_x^3 u + \mathcal{O}(\Delta x^4) ,$$

127) For a thorough treatment, see R.D. Richtmyer, K.W. Morton: *Difference methods for initial-value problems, 2nd ed.* New York: Interscience 1967

into the upwind scheme results in

$$\partial_t u + \frac{\Delta t}{2}\partial_t^2 u + \frac{\Delta t^2}{6}\partial_t^3 u + \mathcal{O}(\Delta t^3) = -a\partial_x u + a\frac{\Delta x}{2}\partial_x^2 u - a\frac{\Delta x^2}{6}\partial_x^3 u + \mathcal{O}(\Delta x^3)\ ,$$

or, after rearranging terms,

$$\partial_t u + a\partial_x u = -\frac{\Delta t}{2}\partial_t^2 u + \frac{a\Delta x}{2}\partial_x^2 u - \frac{\Delta t^2}{6}\partial_t^3 u - \frac{a\Delta x^2}{6}\partial_x^3 u + \mathcal{O}(\Delta t^3, \Delta x^3)\ . \tag{7.29}$$

We now want to replace the time derivatives on the right hand side by space derivatives. In order to do so, we are *not* allowed to use the differential equation (7.28), since our u does not necessarily satisfy this equation, but Eq. (7.29)! Therefore, we must accomplish our task using (7.29) alone. We take the time derivative of (7.29), i.e.,

$$\partial_t^2 u + a\partial_t\partial_x u = -\frac{\Delta t}{2}\partial_t^3 u + \frac{a\Delta x}{2}\partial_t\partial_x^2 u - \frac{\Delta t^2}{6}\partial_t^4 u + \frac{a\Delta x^2}{6}\partial_t\partial_x^3 u + \mathcal{O}(\Delta t^3, \Delta x^3)\ ,$$

and then the space derivative of (7.29), which we multiply by a,

$$a\partial_t\partial_x u + a^2\partial_x^2 u = -a\frac{\Delta t}{2}\partial_t^2\partial_x u + \frac{a^2\Delta x}{2}\partial_x^3 u - \frac{a\Delta t^2}{6}\partial_t^3\partial_x u - \frac{a^2\Delta x^2}{6}\partial_x^4 u + \mathcal{O}(\Delta t^3, \Delta x^3)\ .$$

Subtracting the last equation from the first one gives

$$\partial_t^2 u = a^2\partial_x^2 u + \Delta t\left(-\frac{1}{2}\partial_t^3 u + \frac{a}{2}\partial_t^2\partial_x u\right) + \Delta x\left(\frac{a}{2}\partial_t\partial_x^2 u - \frac{a^2}{2}\partial_x^3 u\right) + \mathcal{O}(\Delta t^3, \Delta x^3)\ .$$

Some other, similar, manipulations of (7.29) lead to

$$\begin{aligned}
\partial_t^3 u &= -a^3\partial_x^3 u + \mathcal{O}(\Delta t, \Delta x)\ ,\\
\partial_x\partial_t^2 u &= a^2\partial_x^3 u + \mathcal{O}(\Delta t, \Delta x)\ ,\\
\partial_t\partial_x^2 u &= -a\partial_x^3 u + \mathcal{O}(\Delta t, \Delta x)\ .
\end{aligned}$$

Substituting these expressions into (7.29) yields the *modified equation*

$$\partial_t u + a\partial_x u = \frac{a\Delta x}{2}\left(1 - \frac{\Delta t}{\Delta x}\right)\partial_x^2 u - \frac{a\Delta x^2}{6}\left(2\frac{\Delta t^2}{\Delta x^2} - 3\frac{\Delta t}{\Delta x} + 1\right)\partial_x^3 u + \mathcal{O}(\Delta x^3, \Delta x^2\Delta t, \Delta x\Delta t^2, \Delta t^3)\ .$$

What can be seen from this equation? First of all, we see that the truncation error of our difference scheme applied to transport equation (7.28) is $\mathcal{O}(\Delta t, \Delta x)$, since the first term on the right side of the modified equation is

$$\frac{a}{2}\Delta x \partial_x^2 u - \frac{a}{2}\Delta t \partial_x^2 u$$

which goes to zero linearly as time and space increase. Moreover, our upwind scheme is a good approximation to the parabolic differential equation

$$\partial_t u + a\partial_x u = \varepsilon(a, \Delta t)\partial_x^2 u$$

with a positive (due to the CFL condition) diffusion coefficient. Since such equations have smooth solutions, we expect our upwind scheme to be well behaved.

Many further properties of the difference scheme can be deduced from the modified equation, e.g., numerical stability.

7.7 Difference Schemes in Conservation Form

It seems obvious that numerical procedures can only be expected to lead to good approximations of the solution to the original problem if the scheme imitates as many of the properties known in advance as possible.

One of these reproductions, already mentioned in Chapter 6, is the use of so-called methods of conservation form in order to treat conservation law equations numerically.

This means that the system of conservation laws (2.1), i.e.,

$$\partial_t \boldsymbol{V} + \partial_x \boldsymbol{f}(\boldsymbol{V}) = \partial_t \boldsymbol{V} + \boldsymbol{Jf}(\boldsymbol{V})\,\partial_x \boldsymbol{V} = 0 \tag{7.30}$$

is approximated by an explicit $(2k+1)$-step finite difference scheme

$$\frac{1}{\Delta t}\left(\boldsymbol{V}_j^{n+1} - \boldsymbol{V}_j^n\right) + \frac{1}{\Delta x}\left(\boldsymbol{g}_{j+\frac{1}{2}}^n - \boldsymbol{g}_{j-\frac{1}{2}}^n\right) = 0 \tag{7.31}$$

with a consistent numerical flux

$$\boldsymbol{g}_{j+\frac{1}{2}}^n := \boldsymbol{g}\left(\boldsymbol{V}_{j-k+1}^n, \boldsymbol{V}_{j-k}^n, \ldots, \boldsymbol{V}_j^n, \boldsymbol{V}_{j+1}^n, \ldots, \boldsymbol{V}_{j+k}^n\right) \tag{7.32}$$

and not (for example) by a method like

$$\frac{1}{\Delta t}\left(\boldsymbol{V}_j^{n+1} - \boldsymbol{V}_j^n\right) + \boldsymbol{Jf}(\boldsymbol{V}_j^n)\frac{1}{2\,\Delta x}\left(\boldsymbol{V}_{j+1}^n - \boldsymbol{V}_{j-1}^n\right) = 0\,.$$

An example of a method of type (7.31) is the three-point *Lax–Friedrichs scheme*[128)]

$$\boldsymbol{V}_j^{n+1} - \frac{1}{2}\left(\boldsymbol{V}_{j+1}^n + \boldsymbol{V}_{j-1}^n\right) + \frac{1}{2}\frac{\Delta t}{\Delta x}\left\{\boldsymbol{f}(\boldsymbol{V}_{j+1}^n) - \boldsymbol{f}(\boldsymbol{V}_{j-1}^n)\right\} = 0 \tag{7.33}$$

128) K.O. Friedrichs, P.D. Lax: Proc. Nat. Acad. Sci. USA **68** (1971) 1686–1688

with

$$g^n_{j+\frac{1}{2}} = -\frac{1}{2}\frac{\Delta x}{\Delta t}\left(V^n_{j+1} - V^n_j\right) + \frac{1}{2}\left\{f(V^n_{j+1}) + f(V^n_j)\right\} .$$

If the numerical vectors V^n_j in (7.33) are replaced by an exact smooth solution $V(x_j, t_n)$ from (7.30), the left side of (7.33) differs from zero, but its Taylor expansion then leads to a truncation error of order $\mathcal{O}(\Delta t)$ in the sense of the Lax–Richtmyer theory, provided that $\lambda = \frac{\Delta t}{\Delta x} = const$ is prescribed. Hence, (7.33) is a method of conservation form of order one. It can be shown that the method is monotone according to the meaning given in Section 6.3, and, as a matter of fact, *all monotone methods are of order one*, as can be established using the "modified equation" in Section 7.6.

However, there are of course also methods of higher order. One of them is the three-point *Lax–Wendroff scheme.*[129] For scalar one-dimensional conservation laws of type (2.1), its numerical flux is given by

$$g^n_{j+\frac{1}{2}} = \frac{1}{2}\left\{f\left(v^n_{j+1}\right) + f\left(v^n_j\right) - \frac{1}{\lambda}\left(\mu^n_{j+\frac{1}{2}}\right)^2\left(v^n_{j+1} - v^n_j\right)\right\} \tag{7.34}$$

with

$$\mu^n_{j+\frac{1}{2}} = \lambda\frac{f(v^n_{j+1}) - f(v^n_j)}{v^n_{j+1} - v^n_j}, \quad \lambda = \frac{\Delta t}{\Delta x} .$$

By Taylor expansion, it can be verified that this is a method of order two. Therefore, the scheme seems to be a useful one, at least as long as the smoothness assumptions necessary for the Taylor expansion are guaranteed in the neighborhood of (x_j, t_n).

However, if the exact weak solution shows a shock at a certain position, the numerical solution begins to oscillate close to this position, although the scheme respects the conservational character of the original problem.

What is not respected by the method is the TVD property of the solutions to the original problem.

Numerical solutions produced via TVD methods do not begin to oscillate, because this would lead to an increase in the total variation.

Several attempts were made in the literature to overcome the problem of low monotone method order and retain the advantages of higher order methods, e.g., the use of greater step sizes without any loss of accuracy in areas where the exact solution is smooth.

One of these ideas, e.g., for scalar problems, involves constructing so-called *flux limiter methods.*

Let g_H be the numerical flux of a higher order method, e.g., of the Lax–Wendroff scheme, and let g_L be the numerical flux of a low-order procedure, e.g., a TVD scheme.

129) P.D. Lax, B. Wendroff: Comm. Pure Appl. Math. **13** (1960) 217–237

If $g_H(v^n_{j+k-1}, \dots v^n_{j+k})$ is abbreviated by $g_H(v^n;j)$ and analogously $g_L(v^n;j)$, let us create a new method via

$$g(v^n;j) := g_L(v^n;j) + \varphi(v^n;j)\,\{g_H(v^n;j) - g_L(v^n;j)\} \ . \tag{7.35}$$

Here, φ – called the *limiter* – must be chosen in such a way that $\varphi \approx 1$ in areas where the exact entropy solution is expected to be smooth, and $\varphi \approx 0$ close to shocks. In other words, φ has to be a sensor for sudden strong increases or decreases in the unknown exact solution that are expected to occur at positions where the numerical solution increases or decreases strongly with respect to space, e.g.,

$$\varphi = \varphi\left(\frac{v^n_j - v^n_{j-1}}{v^n_{j+1} - v^n_j}\right) \tag{7.36}$$

where φ is a sufficiently smooth and bounded function with $\varphi(1) = 1$ and $\varphi(0) = 0$.

Of course, difficulties can still cause trouble, e.g., if an extremum of the solution lies between x_j and x_{j+1}.

Other ideas have also been devised to surmount the problem of nonphysical oscillations when higher-order methods are being considered. One example used for higher dimensions are the so-called *ENO schemes* (**e**ssentially **n**on-**o**scillatory schemes).[130] This idea involves piecewise polynomial reconstruction from the piecewise constant values of the approximate solution for the last time step, where the stencil for each cell is chosen adaptively in such a way that intensive oscillations do not occur near shocks.

7.8 The Finite Volume Method on Unstructured Grids

Let us finally briefly put the idea of finite volume schemes on unstructured grids, introduced in Section 7.1, into concrete form. For convenience, we restrict ourselves to the scalar two-dimensional conservation law (cf. (1.10))

$$\begin{aligned} \partial_t u + \partial_x f_1(u) + \partial_y f_2(u) &= 0 \quad \text{in} \quad \mathbb{R}^2 \times \mathbb{R}^+ , \\ u(x, y; 0) &= u_0(x, y) \quad \text{in} \quad \mathbb{R}^2 . \end{aligned} \tag{7.37}$$

A so-called *unstructured grid* of $\hat{\Omega} \subset \mathbb{R}^2$ consists of a set of polygons σ_i $(i \in I \subset \mathbb{N})$, e.g., triangles, each of which has the same number m of edges and vertices, so that

$$\hat{\Omega} = \bigcup_{i \in I} \sigma_i \ ,$$

and two polygons are not assumed to intersect, or to only intersect along a common edge or at a common vertex. These polygons are often called *cells*. The angles at the

130) See A. Harten: *Multi-dimensional ENO schemes for general geometries (ICASE Report 91-76)*. Hampton: Langley Research Center 1991

vertices must be bounded uniformly away from zero, and we assume $h = \mathcal{O}(\Delta t)$, where h represents a measure of the maximal diameter of the cells.

Unstructured grids can be more easily adapted to complicated geometries than structured grids (e.g., to the gas flow around flying objects), and refinements of the net in regions where the exact solution is expected to vary more than in neighboring areas are easily realized. The development of *grid generation techniques* is therefore a very important task in this context.

Let $I_i \subset I$ be the index set consisting of the numbers of the particular cells that are neighbors of σ_i with a joint edge.

With $u^n(x, y) := u(x, y; n\,\Delta t)$ $(n \in \mathbb{N}_0)$ in the case of a given time step Δt, and if an exact solution u is assumed to be smooth for a short time, Eq. (7.37) leads to

$$\frac{u^{n+1}(x, y) - u^n(x, y)}{\Delta t} + \partial_x f_1(u^n(x, y)) + \partial_y f_2(u^n(x, y)) = \mathcal{O}(\Delta t)\ . \tag{7.38}$$

Taking the divergence theorem into account, integration of this equation over the cell σ_i with edges $\partial\sigma_{i,k}$ $(k = 1, 2, \ldots, m)$ results in

$$\frac{1}{\Delta t}\int_{\sigma_i}\left\{u^{n+1}(x, y) - u^n(x, y)\right\}\,\mathrm{d}\sigma_i + \sum_{k\in I_i}\int_{\partial\sigma_{i,k}}\left\langle \boldsymbol{f}\left(u^n\right), \boldsymbol{n}^{i,k}\right\rangle\,\mathrm{d}s = \mathcal{O}(\Delta t) \tag{7.39}$$

where $\boldsymbol{f}(u) = \left(f_1(u), f_2(u)\right)^T$ is assumed to be sufficiently smooth, and where $\boldsymbol{n}^{i,k}$ denotes the outer normal unit vector on $\partial\sigma_{i,k}$.[131)][132)]

If we replace the functions $u^n(x, y)$, $u^{n+1}(x, y)$ along the cell σ_i by constant approximate values v_i^n, v_i^{n+1} (to be calculated), if the term $\mathcal{O}(\Delta t)$ is neglected, and if $|\sigma_i|$ denotes the area of σ_i and $s_{i,k}$ the length of the edge $\partial\sigma_{i,k}$, (7.39) leads to

$$v_i^{n+1} = v_i^n - \frac{\Delta t}{|\sigma_i|}\sum_{k\in I_i}\left\langle \boldsymbol{f}\left(\tilde{v}_{i,k}^n\right), \boldsymbol{n}^{i,k}\right\rangle\, s_{i,k} \tag{7.40}$$

$(n = 0, 1, 2, \ldots)$, where the values $\tilde{v}_{i,k}^n$ represent approximations for u along the edges $\partial\sigma_{i,k}$, respectively, and where v_i^0 is chosen to be the mean value of the initial function on σ_i:

$$v_i^0 = \frac{1}{|\sigma_i|}\int_{\sigma_i} u_0(x, y)\,\mathrm{d}\sigma_i\ , \qquad \forall\, i \in I\ .$$

Obviously, from a physical point of view, the term $\langle \boldsymbol{f}(\tilde{v}_{i,k}^n), \boldsymbol{n}^{i,k}\rangle\, s_{i,k}$ approximately describes the flux through the edge $\partial\sigma_{i,k}$.

The question of how the values $\tilde{v}_{i,k}^n$ should be chosen then arises.

Because the conservation principle should also be reflected by the approximate values, we aim for

$$\left\langle \boldsymbol{f}\left(\tilde{v}_{i,k}^n\right), \boldsymbol{n}^{i,k}\right\rangle\, s_{i,k} \approx -\left\langle \boldsymbol{f}\left(\tilde{v}_{k,i}^n\right), \boldsymbol{n}^{k,i}\right\rangle\, s_{k,i} \quad (s_{k,i} = s_{i,k})\ . \tag{7.41}$$

131) See (7.4)

132) Here, we assume that each of the cells has neighbors along each of its edges. If this is not the case, namely parts of the boundary of $\hat{\Omega}$ are along the edges, appropriate boundary conditions must be respected.

In order to create a scheme that is easy to handle, we are going to replace $\tilde{v}^n_{i,k}$ and $\tilde{v}^n_{k,i}$ by a certain mean value of v^n_i and v^n_k and replace the physical flux with a numerical flux $g_{i,k}(v^n_i, v^n_k)$ that fulfills relation (7.41) exactly, i.e.,

$$g_{i,k}(v^n_i, v^n_k) = -g_{k,i}(v^n_k, v^n_i) \,. \tag{7.42}$$

Moreover,

$$g_{i,k}(w, w) = \langle \boldsymbol{f}(w), \boldsymbol{n}^{i,k} \rangle \, s_{i,k} \tag{7.43}$$

should be fulfilled, making the finite volume method that is now constructed, namely

$$v^{n+1}_i = v^n_i - \frac{\Delta t}{|\sigma_i|} \sum_{k \in I_i} g_{i,k}(v^n_i, v^n_k) \,, \tag{7.44}$$

consistent with (7.39). Finally, we now forget about the assumption regarding the smoothness of the exact solution, because the integration process also allows us to deal with weak solutions.

Convergence proofs for FVMs in higher dimensions and on unstructured grids, particularly mathematical proofs for convergence to the entropy solutions, are often not currently available. However, it is possible to formulate the methods in such a way that they lead to well-known methods that work well if analogously formulated for one-dimensional problems.

One method that can be shown to behave very well in the 1D situation is the Engquist–Osher flux splitting scheme[133)] with a numerical flux described by Eq. (6.24).

If the numerical flux $g_{i,k}$ in (7.43) is chosen in accord with (6.24) to be

$$g_{i,k}(v, \hat{v}) = \left[\hat{f}^-_{i,k}(v) + \hat{f}^+_{i,k}(\hat{v}) \right] \tag{7.45}$$

with

$$\hat{f}_{i,k}(w) := \langle \boldsymbol{f}(w, \boldsymbol{n}^{i,k} \rangle \, s_{i,k}$$

and with

$$\begin{aligned} \hat{f}^+_{i,k}(v) &= \int_0^v \max\left\{ \hat{f}'_{i,k}(w), 0 \right\} \, dw + \hat{f}_{i,k}(0) \\ \hat{f}^-_{i,k}(\hat{v}) &= \int_0^{\hat{v}} \min\left\{ \hat{f}'_{i,k}(w), 0 \right\} \, dw \,, \end{aligned} \tag{7.46}$$

it corresponds to the Engquist–Osher scheme in 1D, and it fulfills the relations (7.42) and (7.43).

133) See Section 6.3

7.9 Continuous Convergence of Relations

Most of the convergence concepts for discretization procedures mentioned within this book can be considered realizations of an easy general concept called *continuous convergence of relations.*

The convergence results for the Lax–Richtmyer theory (cf. Theorem 7.1) and the convergence proofs in connection with Rinow's theorem (cf. Theorem 6.3) or with the general considerations of Section 6.3 can be regarded as concretizations of this concept, which is based on the ideas of P. du Bois-Reymond, R. Courant and W. Rinow already mentioned in footnotes 103–105.

Roughly speaking, the continuous convergence of objects P_n acting on elements u^n $(n = 1, 2, \ldots)$ to an object P acting on an element u implies

$$u^n \to u \wedge P_n u^n \to v \Rightarrow Pu = v \,. \tag{7.47}$$

We write

$$P_n \xrightarrow{c} P \,.$$[134]

The general concept reads as follows:[135]

Assume that there is a metric space $\mathcal{V}$, an index set $\mathcal{J}$ and a relation R such that, for ordered pairs $(u, \Phi) \in \mathcal{V} \times \mathcal{J}$, the question of whether (u, Φ) belongs to $R \subset \mathcal{V} \times \mathcal{J}$ or not can be answered. If the relation is fulfilled, i.e., if $(u, \Phi) \in R$, we write

$$uR\Phi \,.$$

Moreover, asume that there are subsets $\mathcal{V}_n \subset \mathcal{V}$ $(n = 1, 2, \ldots)$ and relations R_n concerning ordered pairs $(u^n, \Phi) \in \mathcal{V}_n \times \mathcal{J}$ $(n = 1, 2, \ldots)$ such that the question of whether or not u^n is related to Φ arises. If the answer is "yes," i.e., if $(u^n, \Phi) \in R_n$, we write

$$u^n R_n \Phi \,.$$

Definition

We call the sequence $\{R_n\}$ of relations *continuously convergent* to the relation R with respect to the triple $(\mathcal{V}, \{\mathcal{V}_n\}, \mathcal{J})$ if the following implication holds:

$$\forall \Phi \in \mathcal{J} : \; \{u^n | u^n \in \mathcal{V}_n \; (n = 1, 2, \ldots) \; ; \; u^n R_n \Phi\} \to u \Rightarrow uR\Phi \,. \tag{7.48}$$

We then write

$$R_n \xrightarrow{c} R \,.$$

134) This is a little more general than the definition of continuously convergent sequences of operators mentioned in Section 6.3.

135) R. Ansorge: ZAMM **86** (2006) 656–664

136) Numerical equations which do not have a solution do not make sense.

Considering our focus on weakly formulated problems that do not necessarily have unique solutions, it could be interesting to see how this general concept fits with this situation, which is the situation discussed in Sections 6.2 and 6.3.

Assume that there is a given problem

$$Au = v \tag{7.49}$$

where A maps a metric space $\mathcal{V}$ into a metric space $\mathcal{W}$ with a given right hand side $v \in \mathcal{W}$. We ask for solutions $u \in \mathcal{V}$ of problem (7.49). Let

$$S = \{u \in \mathcal{V} | u \text{ solves (7.49)}\} \ .$$

Provided that S contains more than one element, we ask for a unique entropy solution $u_{\mathrm{E}} \in S$. For this reason, we take the ordered pairs $\{(u, \Phi)| u \in S,\ \Phi \in \mathcal{V}\}$ into account where $\mathcal{J}$ is a suitable index set, and let R be a relation between elements of S and $\mathcal{J}$ so that there is *at most* one element in S, denoted by u_{E} (if it exists), which fulfills the entropy condition

$$u_{\mathrm{E}} R \Phi \ , \quad \forall \Phi \in \mathcal{J} \ . \tag{7.50}$$

Hence, the complete problem to be solved consists of finding a solution u_{E} to problem (7.49) which additionally fulfills the relation (7.50).

Let $\mathcal{V}_n \subset \mathcal{V}$ $(n = 1, 2, \ldots)$ be a sequence of subspaces and let $\{v^n | v^n \in \mathcal{V}_n, (n = 1, 2, \ldots)\} \subset \mathcal{V}$ be a given sequence with

$$\lim_{n \to \infty} v^n = v \ . \tag{7.51}$$

The numerical procedure for solving (7.49) approximately then consists of constructing operators A_n: $\mathcal{V}_n \to \mathcal{W}$, $(n = 1, 2, \ldots)$ such that each of the problems

$$A_n u^n = v^n \ , \quad (n = 1, 2, \ldots) \tag{7.52}$$

has at least one solution,[136] i.e.,

$$S_n := \{u^n \in \mathcal{V}_n | u^n \text{ solves (7.52)}\} \neq \emptyset$$

for each fixed $n \in \mathcal{N}$.

Here, it can happen that S_n contains more than one element, but let

$$S_n \to S$$

in the sense of the set convergence described by the last definition in Section 6.2.

Then the following theorem can be stated:

Theorem 7.2

For each fixed $n \in \mathcal{N}$, let R_n be a relation between elements of S_n and of $\mathcal{J}$, and assume that

- There is at least one element $u_{\mathrm{E}}^n \in S_n$ with $u_{\mathrm{E}}^n R_n \Phi$, $\quad \forall \Phi \in \mathcal{J}$, $(n = 1, 2, \ldots)$
- $R_n \xrightarrow{c} R$.

Then there is a unique entropy solution u_{E} to the problem $\{(7.49), (7.50)\}$, and each full sequence

$$\{u_{\mathrm{E}}^n | u_{\mathrm{E}}^n \in S_n\,, \quad (n = 1, 2, \ldots)\} \tag{7.53}$$

converges to u_{E}.

Proof:

Because of the assumptions, there is at least one sequence of type (7.53), and each of these sequences is of the type $\{u^n | u^n \in S_n, (n = 1, 2, \ldots)\}$. Take one of the convergent subsequences of a particular type-(7.53) sequence and denote its limit as u_{E}. $S_n \rightarrow S$ implies $u_{\mathrm{E}} \in S$, and $R_n \xrightarrow{c} R$ yields

$$u_{\mathrm{E}} R \Phi\,, \quad \forall \Phi \in \mathcal{J}\,.$$

However, because there is at most one element $u_{\mathrm{E}} \in S$ which fulfills (7.49), u_{E} is unique and depends on neither the particular type-(7.53) sequence nor the special convergent subsequence of this sequence. Also, because of the uniqueness of u_{E}, not only subsequences but also the full sequence converge to this limit.

It is easy to identify the abstract relations used in the terminology of this section with the different concrete mathematical objects of previous convergence results. However, the concept of continuous convergence of relations also applies to other fields of applied mathematics, e.g., to approximation theory.

8
A Closer Look at Discrete Models

8.1
The Viscosity Form

We have already briefly discussed Riemann problems and their meaning for numerical purposes in Chapter 4. Now we shall identify the Riemann problem as one of the central issues of discrete models of hyperbolic conservation laws.

We start with a discussion of the scalar case[137] and then turn towards the Riemann problem for the Euler equations.

The building block for any finite difference model of scalar conservation laws

$$\partial_t u + \partial_x f(u) = 0$$

with initial conditions $u(x, 0) = u_0(x)$ is a robust first-order numerical flux function $g^n_{j+\frac{1}{2}} = g(v^n_{j+1}, v^n_j)$ to be used in the conservation form

$$v^{n+1}_j = v^n_j - \frac{\Delta t}{\Delta x}\left(g^n_{j+\frac{1}{2}} - g^n_{j-\frac{1}{2}}\right), \tag{8.1}$$

cf. (6.14). In contrast to (6.15), we derive a theory for three-point numerical fluxes, since three points are sufficient for first-order accuracy. We have already encountered the first-order monotone *Lax–Friedrichs scheme* (7.33) defined through the numerical flux

$$g^n_{j+\frac{1}{2}} := \frac{\Delta x}{2\Delta t}\left(v^n_{j+1} - v^n_j\right) + \frac{1}{2}\left\{f(v^n_{j+1}) + f(v^n_j)\right\}. \tag{8.2}$$

The form in which this numerical flux has been written is not very instructive. We will therefore exploit the idea of considering a discrete model to be a discretization of the perturbed equation

$$\partial_t u + \partial_x f(u) = \partial_x(Q(u)\partial_x u). \tag{8.3}$$

The right hand side perturbation models a stabilizing diffusion term in which the function $u \mapsto Q(u)$ is called the *viscosity coefficient*. It is one of the most important

137) See E. Tadmor: Math. Comp. **43** (1984) 353–368; E. Tadmor: Math. Comp. **43** (1984) 369–381

Mathematical Models of Fluid Dynamics. R. Ansorge and T. Sonar

ISBN: 978-3-527-40774-3

and striking consequences of the theory we shall study that some sort of viscosity is necessary in order to build stable discrete models. With (8.3) in mind, a discrete numerical flux should always have the *viscosity form*

$$g^n_{j+\frac{1}{2}} = \frac{1}{2}\left(f(v^n_j) + f(v^n_{j+1})\right) + \frac{\Delta x}{2\Delta t} Q^n_{j+\frac{1}{2}}(v^n_{j+1} - v^n_j) \tag{8.4}$$

with a *numerical viscosity coefficient* $Q^n_{j+\frac{1}{2}} = Q(v^n_j, v^n_{j+1})$. As a consequence, the difference

$$g^n_{j+\frac{1}{2}} - g^n_{j-\frac{1}{2}} = \frac{1}{2}\left(f(v^n_{j+1}) - f(v^n_j)\right) + \frac{\Delta x}{2\Delta t}\left\{Q^n_{j+\frac{1}{2}} \cdot (v^n_{j+1} - v^n_j) - Q^n_{j-\frac{1}{2}} \cdot (v^n_j - v^n_{j-1})\right\}$$

clearly shows the reason for using (8.3) as a model equation. The case $Q^n_{j+\frac{1}{2}} \equiv 0$, i.e., no viscosity, results in the central difference of the fluxes, which is – after linearization – shown to be unconditionally unstable in von Neumann's stability theory (cf. Section 7.5).

One could now argue that any positive function Q (however small) could lead to stable schemes, but – as we shall see – this is by no means true!

8.2 The Incremental Form

In Section 6.3 we introduced TVD and monotone methods of type (8.1). Method (8.1) is said to be TVD if

$$\forall n \in \mathbb{N}: \quad \sum_j \left| v^{n+1}_{j+1} - v^{n+1}_j \right| \le \sum_j \left| v^n_{j+1} - v^n_j \right| .$$

We call methods of type (8.1) *monotonicity preserving* if

$$v^n_{\cdot} \text{ monotone} \Rightarrow v^{n+1}_{\cdot} \text{ monotone} \tag{8.5}$$

holds. We can now easily prove that

$$\text{(8.1) monotone} \Rightarrow \text{(8.1) TVD} \Rightarrow \text{(8.1) monotonicity preserving} \tag{8.6}$$

holds. One can easily show that solutions to monotone schemes form l^1-contractive semigroups, i.e., that

$$\|v^{n+1}_{\cdot+1} - v^{n+1}_{\cdot}\|_{l^1(\mathbb{R})} \le \|v^n_{\cdot+1} - v^n_{\cdot}\|_{l^1(\mathbb{R})}$$

is valid, where $\|v^n_{\cdot}\|_{l^1(\mathbb{R})} := \sum_j |v^n_j|$. Hence, the first implication follows. To prove the second, we consider the grid function

$$\begin{aligned} v^n_j &= \text{const} =: v^n_L\,; \quad j < k-1 \\ v^n_j & \quad \text{monotonically increasing}\,; \quad k-1 \le j \le k+1 \\ v^n_j &= \text{const} =: v^n_R\,; \quad j > k+1\,. \end{aligned}$$

In that case, the total variation is $|v^n_R - v^n_L|$. Assume that $v^{n+1}_{\cdot}$ does not monotonically increase, so that we have a maximum and a minimum – say v^{n+1}_M and v^{n+1}_m, respectively. It would follow that

$$TV(v_{\cdot}^{n+1}) := \sum_j |v_{j+1}^{n+1} - v_j^{n+1}| \geq |v_R^n - v_L^n| + |v_M^{n+1} - v_m^{n+1}| > TV(v_{\cdot}^n) ,$$

which contradicts the TVD property. This concludes the proof.

The *incremental form* of a scheme (8.1) can be achieved if there are incremental factors $C_{j+\frac{1}{2}}^{\pm}$, so that (8.1) can be written as

$$v_j^{n+1} = v_j^n + C_{j+\frac{1}{2}}^+ \cdot (v_{j+1}^n - v_j^n) - C_{j-\frac{1}{2}}^- \cdot (v_j^n - v_{j-1}^n) . \tag{8.7}$$

It is an easy exercise to show that every scheme (8.1) can be put into incremental form, and that the incremental factors are given by

$$C_{j+\frac{1}{2}}^+ = \frac{\Delta t}{\Delta x} \frac{f(v_j^n) - g_{j+\frac{1}{2}}^n}{v_{j+1}^n - v_j^n}$$

$$C_{j-\frac{1}{2}}^- = \frac{\Delta t}{\Delta x} \frac{f(v_j^n) - g_{j-\frac{1}{2}}^n}{v_j^n - v_{j-1}^n} .$$

Theorem 8.1

Monotonicity-preserving three-point schemes satisfying

$$C_{j+\frac{1}{2}}^+ \geq 0 ; \quad C_{j+\frac{1}{2}}^- \geq 0 ; \quad 1 - C_{j+\frac{1}{2}}^+ - C_{j+\frac{1}{2}}^- \geq 0 \tag{8.8}$$

are TVD schemes.

Proof:

Applying difference operators Δ to (8.7) yields

$$\Delta v_j^n = C_{j+\frac{1}{2}}^+ \Delta v_{j+1}^n + \left(1 - C_{j+\frac{1}{2}}^+ - C_{j+\frac{1}{2}}^-\right) \Delta v_j^n + C_{j-\frac{1}{2}}^- \Delta v_{j-1}^n . \tag{8.9}$$

In the case of $\Delta v_{j+1}^n = \Delta v_{j-1}^n = 0$, it follows that

$$\Delta v_j^{n+1} = \left(1 - C_{j+\frac{1}{2}}^+ - C_{j+\frac{1}{2}}^-\right) \Delta v_j^n .$$

Preserving monotonicity means sign Δv_j^{n+1} = sign Δv_j^n, and hence

$$1 - C_{j+\frac{1}{2}}^+ - C_{j+\frac{1}{2}}^- \geq 0 .$$

The two remaining inequalities follow analogously. Summing (8.9) yields

$$TV(v_{\cdot}^{n+1}) \leq \sum_j |C_{j+\frac{3}{2}}^+||\Delta v_{j+1}^n| + \sum_j |1 - C_{j+\frac{1}{2}}^+ - C_{j+\frac{1}{2}}^-||\Delta v_j^n| + \sum_j |C_{j-\frac{1}{2}}^-||\Delta v_{j-1}^n|$$

and an index shift in the first sum and third sum reveals $TV(v_{\cdot}^{n+1}) \leq TV(v_{\cdot}^n)$.

8.3 Relations

With an eye on (8.1) with (8.4), as well as on (8.7), we can identify the numerical viscosity coefficient as

$$Q^n_{j+\frac{1}{2}} = C^+_{j+\frac{1}{2}} + C^-_{j+\frac{1}{2}} = \frac{\Delta t}{\Delta x} \frac{f(v^n_j) + f(v^n_{j+1}) - 2g^n_{j+\frac{1}{2}}}{v^n_{j+1} - v^n_j} . \tag{8.10}$$

Conversely, the incremental factors can be represented as

$$C^{\pm}_{j+\frac{1}{2}} = \frac{1}{2}\left(Q^n_{j+\frac{1}{2}} \mp \frac{\Delta t}{\Delta x} \frac{f(v^n_{j+1}) - f(v^n_j)}{v^n_{j+1} - v^n_j} \right) . \tag{8.11}$$

By exploiting relations (8.10) and (8.11), we can already characterize difference schemes (8.1) by their numerical viscosity coefficients.

Theorem 8.2

Difference schemes (8.1) with numerical flux (8.4) are TVD schemes if

$$\frac{\Delta t}{\Delta x} \left| \frac{f(v^n_{j+1}) - f(v^n_j)}{v^n_{j+1} - v^n_j} \right| \le Q^n_{j+\frac{1}{2}} \le 1 \tag{8.12}$$

holds true.

Proof:

If (8.12) holds, then the incremental factors satisfy the assumptions of Theorem 8.1.

Theorem 8.3

If

$$\frac{\Delta t}{\Delta x} \left| \frac{f(v^n_{j+1}) - f(v^n_j)}{v^n_{j+1} - v^n_j} \right| \le Q^n_{j+\frac{1}{2}} \le \frac{1}{2} \tag{8.13}$$

holds, then the maximum principle

$$\forall k \in \mathbb{Z} : \quad \inf_k v^n_k \le v^{n+1}_k \le \sup_k v^n_k \tag{8.14}$$

follows.

Proof:

From (8.7), one derives

$$v^{n+1}_j = C^+_{j+\frac{1}{2}} v^n_{j+1} + (1 - C^+_{j+\frac{1}{2}} - C^-_{j-\frac{1}{2}}) v^n_j + C^-_{j+\frac{1}{2}} v^n_{j-1}$$

which is a convex combination of $v_{j+1}^n, v_j^n, v_{j-1}^n$. Now (8.13) follows because of

$$0 \le C^{\pm}_{j+\frac{1}{2}} = \frac{1}{2}\left(Q^n_{j+\frac{1}{2}} \mp \frac{\Delta t}{\Delta x}\frac{f(v^n_{j+1}) - f(v^n_j)}{v^n_{j+1} - v^n_j}\right) \le \frac{1}{2}\cdot 2Q^n_{j+\frac{1}{2}} \le \frac{1}{2}\ .$$

We have not yet discussed the case when $v^n_{j+1} - v^n_j = 0$. As can be seen from (8.12), in this case the numerical viscosity coefficient has to satisfy the inequality

$$\frac{\Delta t}{\Delta x}|f'(v^n_j)| \le Q^n_{j+\frac{1}{2}} \le 1\ ,$$

or, from (8.13),

$$\frac{\Delta t}{\Delta x}|f'(v^n_j)| \le Q^n_{j+\frac{1}{2}} \le \frac{1}{2}$$

in order to save the maximum principle. This, of course, is simply a type of CFL condition (cf. Section 7.3).

Theorems 8.2 and 8.3 clearly define the range of proper numerical viscosity coefficients, and one could request the two difference schemes that satisfy $Q^n_{j+\frac{1}{2}} \equiv 1$ and $Q^n_{j+\frac{1}{2}} = \frac{\Delta t}{\Delta x}\left|\frac{f(v^n_{j+1}) - f(v^n_j)}{v^n_{j+1} - v^n_j}\right|$, respectively.

These schemes do actually exist, and we have already encountered the one with the highest numerical viscosity possible: setting $Q^n_{j+\frac{1}{2}} \equiv 1$ in (8.4) leads to

$$g^n_{j+\frac{1}{2}} = \frac{1}{2}\left(f(v^n_j) + f(v^n_{j+1})\right) + \frac{\Delta x}{2\Delta t}(v^n_{j+1} - v^n_j)\ ,$$

and this is simply the Lax–Friedrichs flux (8.2), i.e.,

Theorem 8.4

The Lax–Friedrichs scheme is characterized by

$$Q^n_{j+\frac{1}{2}} =: Q^{LF}_{j+\frac{1}{2}} = 1\ . \tag{8.15}$$

8.4 Godunov Is Just Good Enough

Surprisingly, the scheme with the least possible numerical viscosity coefficient has also been known for quite a long time.

In 1952 Courant, Isaacson and Rees published a paper[138] in which they derived the classical Courant–Isaacson–Rees finite difference scheme. Unfortunately, their

138) R. Courant, E. Isaacson, M. Rees: Comm. Pure Appl. Math. **5** (1952) 243–244; see also E.M. Murman: AIAA J. **12** (1974) 626–633 and footnote 87

scheme was not formulated in conservation form, but this was remedied by Murman in 1974. The Murman–Courant–Isaacson–Rees (MCIR) scheme is given by

$$v_j^{n+1} = v_j^n - \frac{\Delta t}{\Delta x}\left(s_{j+\frac{1}{2}}\left(f(v_{j+1}^n) - f(v_j^n)\right) + (1 - s_{j-\frac{1}{2}})\left(f(v_j^n) - f(v_{j-1}^n)\right)\right)$$

where

$$s_{j+\frac{1}{2}} := \frac{1}{2}\left(1 - \operatorname{sign}\left(\frac{f(v_{j+1}^n) - f(v_j^n)}{v_{j+1}^n - v_j^n}\right)\right).$$

It is easily seen that the numerical flux of this difference scheme is given by

$$g_{j+\frac{1}{2}}^n = \frac{1}{2}\left(f(v_j^n) + f(v_{j+1}^n)\right) - \frac{1}{2}\left|\frac{f(v_{j+1}^n) - f(v_j^n)}{v_{j+1}^n - v_j^n}\right| \cdot (v_{j+1}^n - v_j^n),$$

and so the numerical viscosity coefficient is given by

$$Q_{j+\frac{1}{2}}^{\mathrm{MCIR}} := \frac{\Delta t}{\Delta x}\left|\frac{f(v_{j+1}^n) - f(v_j^n)}{v_{j+1}^n - v_j^n}\right| . \tag{8.16}$$

According to Theorem 8.2, this is the least possible numerical viscosity coefficient for a TVD scheme.

There is, however, a serious problem with the MCIR scheme. In the case of Burgers' equation

$$\partial_t u + \partial_x\left(\frac{1}{2}u^2\right) = 0$$

with the initial condition

$$u(x,0) = u_0(x) = \begin{cases} -1\,, & x < 0 \\ 1\,, & x \geq 0 \end{cases},$$

i.e.,

$$v_j^0 = \begin{cases} -1\,, & j < 0 \\ 1\,, & j \geq 0 \end{cases},$$

the scheme gives a stable, entropy-violating stationary shock. The reason for this behavior is easily found: the numerical viscosity coefficient vanishes in the rarefaction shock.

We therefore have to search for a scheme with a nonvanishing viscosity coefficient resembling (8.16). This is the famous *Godunov scheme.*

According to (4.1), we now think of initial values as being piecewise constant on an initial grid, i.e.,

$$v_j^0 := \frac{1}{\Delta x}\int_{\frac{x_{j-1}+x_j}{2}}^{\frac{x_j+x_{j+1}}{2}} u_0(x)\,\mathrm{d}x\,, \quad j \in \mathbb{Z}\,.$$

Hence, the initial values give rise to local Riemann problems at points $\ldots, x_{j-\frac{1}{2}}$, $x_{j+\frac{1}{1}}, \ldots$ Godunov's brillant idea was to solve the local Riemann problems exactly and use the solution as new values at the next time step.

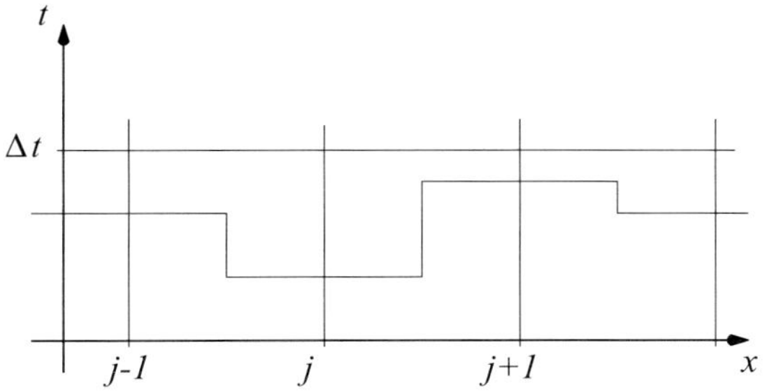

Fig. 8.1 Local Riemann problems.

Definition

Godunov's method is given by the recipe

$$v_j^{n+1} = \frac{1}{2}(u^-_{j-\frac{1}{2}} + u^+_{j+\frac{1}{2}}) \tag{8.17}$$

where

$$u^-_{j-\frac{1}{2}} := \frac{2}{\Delta x} \int\limits_0^{\frac{\Delta x}{2}} u^{\mathrm{R}}(z, t_{n+1}; v^n_{j-1}, v^n_j)\ \mathrm{d}z \tag{8.18}$$

and

$$u^+_{j+\frac{1}{2}} := \frac{2}{\Delta x} \int\limits_{-\frac{\Delta x}{2}}^{0} u^{\mathrm{R}}(z, t_{n+1}; v^n_j, v^n_{j+1})\ \mathrm{d}z\ , \tag{8.19}$$

and

$$u^{\mathrm{R}}(z, t_{n+1}; u_{\mathrm{L}}, u_{\mathrm{R}})$$

denotes the exact solution of the Riemann problem at time t_{n+1} with initial left and right data u_{L} and u_{R}, respectively. Here, z is a local coordinate $z \in [\frac{-\Delta x}{2}, \frac{\Delta x}{2}]$ around $x_{j-\frac{1}{2}}$ and $x_{j+\frac{1}{2}}$, respectively.

In order to avoid interactions between the waves from neighboring Riemann solutions, we have to admit a slightly stronger CFL condition:

$$\frac{\Delta t}{\Delta x} \max_j \left| f'(v^n_j) \right| \leq \frac{1}{2}\ . \tag{8.20}$$

Integrating our differential equation over a "cell" $[x_{j-\frac{1}{2}}, x_{j+\frac{1}{2}}]$ gives

$$\int\limits_{x_{j-\frac{1}{2}}}^{x_{j+\frac{1}{2}}} \int\limits_{t_n}^{t_{n+1}} \partial_t u^{\mathrm{R}}\ \mathrm{d}t\ \mathrm{d}x + \int\limits_{x_{j-\frac{1}{2}}}^{x_{j+\frac{1}{2}}} \int\limits_{t_n}^{t_{n+1}} \partial_x f(u^{\mathrm{R}})\ \mathrm{d}t\ \mathrm{d}x = 0$$

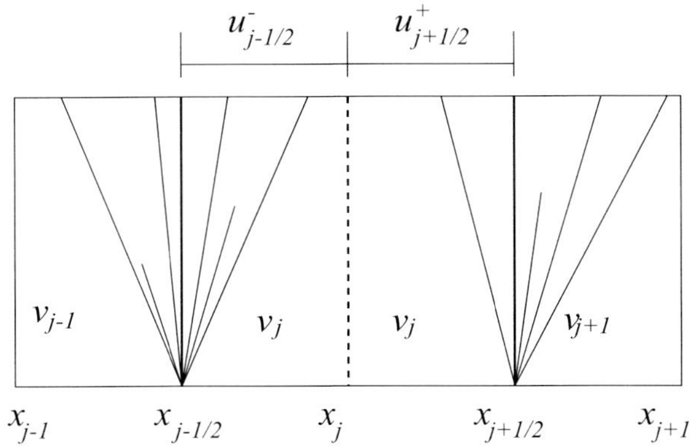

Fig. 8.2 Local Riemann problems.

and hence

$$\int_0^{\frac{\Delta x}{2}} u^{\mathrm{R}}(z, t_{n+1}; v_{j-1}^n, v_j^n)\,\mathrm{d}z + \int_{-\frac{\Delta x}{2}}^0 u^{\mathrm{R}}(z, t_{n+1}; v_j^n, v_{j+1}^n)\,\mathrm{d}z \tag{8.21}$$
$$- \int_0^{\frac{\Delta x}{2}} u^{\mathrm{R}}(z, t_n; v_{j-1}^n, v_j^n)\,\mathrm{d}z - \int_{-\frac{\Delta x}{2}}^0 u^{\mathrm{R}}(z, t_n; v_j^n, v_{j+1}^n)\,\mathrm{d}z$$
$$+ \int_{t_n}^{t_{n+1}} \left\{ f(u^{\mathrm{R}}(0, t; v_j^n, v_{j+1}^n)) - f(u^{\mathrm{R}}(0, t; v_{j-1}^n, v_j^n) \right\} \mathrm{d}t = 0\,.$$

According to (8.17)–(8.19), the sum of the two leading integrals is $\Delta x \cdot v_j^{n+1}$, while the sum of the following two is $\Delta x \cdot v_j^n$. The Riemann solution is constant on rays, so $u^{\mathrm{R}}(0, \ldots)$ is a constant value. Let us denote this value of $u^{\mathrm{R}}(0, t; v_{j-1}^n, v_j^n)$ by $w_{j-\frac{1}{2}}$, and the value $u^{\mathrm{R}}(0, t; v_j^n, v_{j+1}^n)$ by $w_{j+\frac{1}{2}}$. Then the remaining integral in (8.21) evaluates to

$$\Delta t \left(f(w_{j+\frac{1}{2}}) - f(w_{j-\frac{1}{2}}) \right)\,.$$

We have therefore shown in passing that the Godunov difference scheme can be written as

$$v_j^{n+1} = v_j^n - \frac{\Delta t}{\Delta x} \left(g^{\mathrm{G}}_{j+\frac{1}{2}} - g^{\mathrm{G}}_{j-\frac{1}{2}} \right)$$

with the numerical flux

$$g^{\mathrm{G}}_{j+\frac{1}{2}} := f(w_{j+\frac{1}{2}})\,. \tag{8.22}$$

Hence, if there is no state v between v_j^n and v_{j+1}^n for which $f'(v) = 0$, the term $w_{j+\frac{1}{2}}$ will take the value v_{j+1}^n or v_j^n according to where the wave lies. In this case, the numerical Godunov flux can be written as

$$g^{G}_{j+\frac{1}{2}} = \frac{1}{2}\left(f(v^n_j) + f(v^n_{j+1}) - \left|\frac{f(v^n_{j+1}) - f(v^n_j)}{v^n_{j+1} - v^n_j}\right| (v^n_{j+1} - v^n_j)\right) \tag{8.23}$$

and the viscosity coefficient (8.16) of the MCIR scheme can be identified immediately. The Godunov scheme and the MCIR scheme are identical in this case.

In the case of a wave with $f'(u) = 0$, we can get additional information from (8.10). This gives the following formula for the full Godunov viscosity coefficient:

$$Q^{G}_{j+\frac{1}{2}} = \begin{cases} \dfrac{\Delta t}{\Delta x} \dfrac{f(v^n_j) + f(v^n_{j+1}) - 2g^{G}_{j+\frac{1}{2}}}{v^n_{j+1} - v^n_j} & ; \quad f'(v^n_j) < 0 < f'(v^n_{j+1}) \\ \dfrac{\Delta t}{\Delta x} \left|\dfrac{f(v^n_{j+1}) - f(v^n_j)}{v^n_{j+1} - v^n_j}\right| & ; \quad \text{otherwise} \end{cases} \tag{8.24}$$

where we take $g^{G}_{j+\frac{1}{2}}$ from (8.22). Hence, the Godunov scheme has a numerical viscosity coefficient that never vanishes.

Let us now check the behavior of the Godunov scheme with respect to an entropy inequality. As in previous chapters – see Section 3.2 – we denote a convex entropy function by S and a corresponding entropy flux[139] by F. A function u solving a conservation law is called entropy solution if the entropy inequality

$$\frac{\partial S}{\partial t}(u) + \frac{\partial F}{\partial x}(u) \le 0$$

holds for every pair (S, F) in the weak sense. As above, we integrate

$$\int_0^{\frac{\Delta x}{2}} \int_{t_n}^{t_{n+1}} \frac{\partial S}{\partial t}(u^{R})\, dt\, dx + \int_0^{\frac{\Delta x}{2}} \int_{t_n}^{t_{n+1}} \frac{\partial F}{\partial x}(u^{R})\, dt\, dx \le 0$$

and get (cf. (8.21))

$$\int_0^{\frac{\Delta x}{2}} S(u^{R}(z, t_{n+1}; v^n_{j-1}, v^n_j)), \, dz \le \frac{\Delta x}{2} S(v^n_j) - \Delta t \left(F(v^n_j) - G^{G}_{j-\frac{1}{2}}\right)$$

where

$$G^{G}_{j-\frac{1}{2}} = F(w_{j-\frac{1}{2}})$$

is a consistent numerical entropy flux. Completely analogously, we get

$$\int_{-\frac{\Delta x}{2}}^{0} S(u^{R}(z, t_{n+1}; v^n_j, v^n_{j+1}))\, dz \le \frac{\Delta x}{2} S(v^n_j) + \Delta t \left(F(v^n_j) - G^{G}_{j+\frac{1}{2}}\right).$$

139) For the relation between entropy and entropy flux, see (3.13)

140) See, for example, E. Hewitt, K. Stromberg: *Real and abstract analysis*. Berlin: Springer 1965

Since an entropy function is always convex, we can use Jensen's inequality,[140] which gives

$$S(u^-_{j-\frac{1}{2}}) \le \frac{2}{\Delta x} \int_0^{\frac{\Delta x}{2}} S(u^{\mathrm{R}}(z, t_{n+1}; v^n_{j-1}, v^n_j)) \, \mathrm{d}z$$

and

$$S(u^+_{j+\frac{1}{2}}) \le \frac{2}{\Delta x} \int_{-\frac{\Delta x}{2}}^{0} S(u^{\mathrm{R}}(z, t_{n+1}; v^n_j, v^n_{j+1})) \, \mathrm{d}z$$

Employing (8.17) finally yields

$$S(v^{n+1}_j) \le \frac{1}{2} \left(S(u^-_{j-\frac{1}{2}}) + S(u^+_{j+\frac{1}{2}}) \right) \le S(v^n_j) - \frac{\Delta t}{\Delta x} \left(G^{\mathrm{G}}_{j+\frac{1}{2}} - G^{\mathrm{G}}_{j-\frac{1}{2}} \right). \tag{8.25}$$

Hence, Godunovs method satisfies every discrete entropy inequality. We fix our achievements in the following theorem

Theorem 8.5

Under the CFL condition (8.20), it holds that

$$u^-_{j-\frac{1}{2}} = v^n_j - \frac{\Delta t}{\Delta x} \left(f(v^n_j) - f(v^n_{j-1}) \right) - Q^{\mathrm{G}}_{j-\frac{1}{2}} (v^n_j - v^n_{j-1}) \tag{8.26}$$

$$u^+_{j+\frac{1}{2}} = v^n_j - \frac{\Delta t}{\Delta x} \left(f(v^n_{j+1}) - f(v^n_j) \right) + Q^{\mathrm{G}}_{j+\frac{1}{2}} (v^n_{j+1} - v^n_j) \tag{8.27}$$

$$S(u^-_{j-\frac{1}{2}}) \le S(v^n_j) - 2\frac{\Delta t}{\Delta x} \left(F(v^n_j) - G^{\mathrm{G}}_{j-\frac{1}{2}} \right) \tag{8.28}$$

$$S(u^+_{j+\frac{1}{2}}) \le S(v^n_j) + 2\frac{\Delta t}{\Delta x} \left(F(v^n_j) - G^{\mathrm{G}}_{j+\frac{1}{2}} \right) \tag{8.29}$$

Proof:

The entropy inequalities (8.28) and (8.29) have just been proven. The formulae (8.26) and (8.27) can easily be verified.

We have proven that the Godunov scheme is the three-point scheme with the least amount of numerical viscosity possible according to Theorem 8.3. We will see soon that the Lax–Friedrichs scheme – the scheme with maximum possible viscosity – is closely related to the Godunov scheme.

8.5 The Lax–Friedrichs Scheme

We follow Tadmor[141] and rewrite the classical Lax–Friedrichs scheme

$$v^{n+1}_j = \frac{1}{2} \left(v^n_{j+1} + v^n_{j-1} \right) - \frac{\Delta t}{2\Delta x} \left(f(v^n_{j+1}) - f(v^n_{j-1}) \right)$$

141) See footnote 137

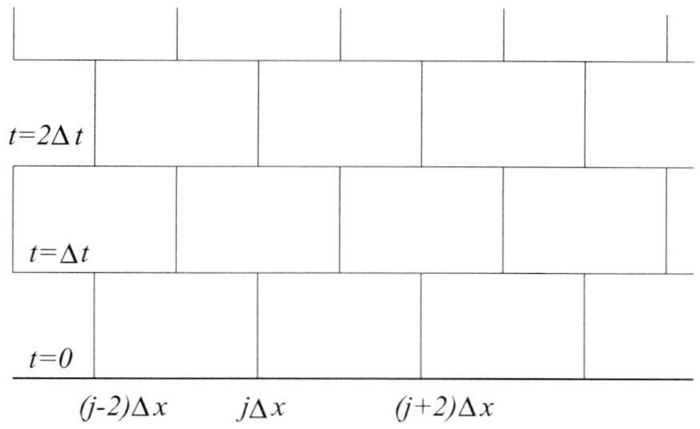

Fig. 8.3 Staggered grid.

with numerical flux

$$g^{\mathrm{LF}}_{j+\frac{1}{2}} = \frac{1}{2}\left(f(v_j^n) + f(v_{j+1}^n) - \frac{\Delta x}{\Delta t}(v_{j+1}^n - v_j^n)\right)$$

and numerical viscosity coefficient

$$Q^{\mathrm{LF}}_{j+\frac{1}{2}} = 1$$

so that it appears as a finite difference scheme on a staggered grid like that shown in Fig. 8.3.

The initial function is now assumed to be constant between $[(j-2)\Delta x, j\Delta x]$, $[j\Delta x, (j+2)\Delta x]$, etc. At the next time step, we must discretize such that we get constant values between $[(j-1)\Delta x, (j+1)\Delta x]$, etc. To avoid wave interactions, we now have to satisfy the CFL condition

$$\frac{\Delta t}{\Delta x}\max_j |f'(v_j^n)| \le 1 ,$$

which is exactly the natural CFL condition for the Lax–Friedrichs scheme. Surprisingly, the following result holds.

Theorem 8.6

The following lemma is known as *Tadmor's lemma*. The Godunov scheme on the staggered grid described above is exactly the Lax–Friedrichs scheme, i.e.,

$$v_j^{n+1} = \frac{1}{2\Delta x}\int_{-\Delta x}^{\Delta x} u^{\mathrm{R}}(z, t_{n+1}; v_{j-1}^n, v_{j+1}^n)\,\mathrm{d}z = \frac{v_{j+1}^n + v_{j-1}^n}{2} - \frac{\Delta t}{2\Delta x}(f(v_{j+1}^n) - f(v_{j-1}^n)) .$$

Proof:

We proceed along the lines already laid out above. Integration of the conservation law gives

$$\int_{-\Delta x}^{\Delta x} u^{\mathrm{R}}(z, t_{n+1}; v_{j-1}^n, v_{j+1}^n)\,\mathrm{d}z - \int_{-\Delta x}^{\Delta x} u^{\mathrm{R}}(z, t_n; v_{j-1}^n, v_{j+1}^n)\,\mathrm{d}z$$
$$+ \int_{t_n}^{t_{n+1}} f(u^{\mathrm{R}}(\Delta x, t; v_{j-1}^n, v_{j+1}^n))\,\mathrm{d}t - \int_{t_n}^{t_{n+1}} f(u^{\mathrm{R}}(-\Delta x, t; v_{j-1}^n, v_{j+1}^n))\,\mathrm{d}t = 0\,.$$

The two time integrals yield $\Delta t(f(v_{j+1}^n) - f(v_{j-1}^n)))$, and the second space integral is by definition just $2\Delta x \cdot v_j^n$. It therefore follows that

$$\int_{-\Delta x}^{\Delta x} u^{\mathrm{R}}(z, t_{n+1}; v_{j-1}^n, v_{j+1}^n)\,\mathrm{d}z = 2\Delta x \cdot v_j^{n+1} = 2\Delta x \cdot v_j^n - \Delta t(f(v_{j+1}^n) - f(v_{j-1}^n))\,,$$

which is the desired result.

The Godunov scheme could be written as the arithmetic mean of two quantities. In order to tailor the Lax–Friedrichs scheme to this form, we introduce a slight modification of the classical Lax–Friedrichs scheme:

$$v_j^{n+1} = \frac{1}{4}\left(v_{j+1}^n + 2v_j^n + v_{j-1}^n\right) - \frac{\Delta t}{2\Delta x}\left(f(v_{j+1}^n) - f(v_{j-1}^n)\right)\,, \tag{8.30}$$

which can be written as

$$v_j^{n+1} = \frac{1}{2}\left(u_{j-\frac{1}{2}}^- + u_{j+\frac{1}{2}}^+\right)\,, \tag{8.31}$$

where

$$u_{j-\frac{1}{2}}^- = \frac{1}{2}(v_j^n + v_{j-1}^n) - \frac{\Delta t}{\Delta x}(f(v_j^n) - f(v_{j-1}^n)) \tag{8.32}$$
$$u_{j+\frac{1}{2}}^+ = \frac{1}{2}(v_{j+1}^n + v_j^n) - \frac{\Delta t}{\Delta x}(f(v_{j+1}^n) - f(v_j^n))\,. \tag{8.33}$$

We call this scheme the *modified Lax–Friedrichs scheme*. Due to the modification, the numerical viscosity coefficient is now

$$Q_{j+\frac{1}{2}}^{\mathrm{mLF}} = \frac{1}{2}\,.$$

According to Tadmor's lemma, we have

$$v_j^{n+1} = \frac{1}{2\Delta x}\int_{-\frac{\Delta x}{2}}^{\frac{\Delta x}{2}} u^{\mathrm{R}}(z, t_{n+1}; v_{j-1}^n, v_j^n)\,\mathrm{d}z + \frac{1}{2\Delta x}\int_{-\frac{\Delta x}{2}}^{\frac{\Delta x}{2}} u^{\mathrm{R}}(z, t_{n+1}; v_j^n, v_{j+1}^n)\,\mathrm{d}z$$

under the CFL condition

$$\frac{\Delta t}{\Delta x}\max_j |f'(v_j^n)| \le \frac{1}{2}\,.$$

In complete analogy to the computations performed for the Godunov scheme, we can verify the validity of a discrete entropy inequality

$$S(v_j^{n+1}) \le S(v_j^n) - \frac{\Delta t}{\Delta x}\left(G_{j+\frac{1}{2}}^{\mathrm{mLF}} - G_{j-\frac{1}{2}}^{\mathrm{mLF}}\right)$$

with numerical entropy flux

$$G_{j+\frac{1}{2}}^{\mathrm{mLF}} = \frac{1}{2}\left(F(v_{j+1}^n) - F(v_j^n)\right) - \frac{\Delta x}{4\Delta t}\left(S(v_{j+1}^n) - S(v_j^n)\right).$$

Hence we have shown:

Theorem 8.7

Under the CFL condition (8.20), it holds that

$$u_{j-\frac{1}{2}}^- = v_j^n - \frac{\Delta t}{\Delta x}\left(f(v_j^n) - f(v_{j-1}^n)\right) - \frac{1}{2}(v_j^n - v_{j-1}^n) \tag{8.34}$$

$$u_{j+\frac{1}{2}}^+ = v_j^n - \frac{\Delta t}{\Delta x}\left(f(v_{j+1}^n) - f(v_j^n)\right) + \frac{1}{2}(v_{j+1}^n - v_j^n) \tag{8.35}$$

$$S(u_{j-\frac{1}{2}}^-) \le S(v_j^n) - 2\frac{\Delta t}{\Delta x}\left(F(v_j^n) - G_{j-\frac{1}{2}}^{\mathrm{mLF}}\right) \tag{8.36}$$

$$S(u_{j+\frac{1}{2}}^+) \le S(v_j^n) + 2\frac{\Delta t}{\Delta x}\left(F(v_j^n) - G_{j+\frac{1}{2}}^{\mathrm{mLF}}\right). \tag{8.37}$$

For the case of scalar conservation laws, we have thus derived a complete theory by going back completely to Tadmor's work. We can now go a step further and develop a theory for all three-point schemes with numerical viscosity coefficient Q satisfying

$$Q_{j+\frac{1}{2}}^{\mathrm{mLF}} \ge Q_{j+\frac{1}{2}} \ge Q_{j+\frac{1}{2}}^{G}\,.$$

This was indeed done,[142] and it leads to a satisfying convergence theory. However, we shall not follow this trail, but instead divert our attention to the Godunov scheme for inviscid fluid flow. Practical experience has revealed that the Godunov scheme is an extremely useful building block for many numerical methods which – at least, if we draw our conclusions from the scalar theory developed above – does not introduce too much numerical viscosity.

142) S. Osher, E. Tadmor: Math. Comp. **50** (1988) 19–51

8.6 A Glimpse of Gas Dynamics

The Riemann problem is at the heart of Godunov's finite difference scheme. We introduced the Riemann problem in Chapter 4, but we now provide a thorough description of it in the case of one-dimensional inviscid flow described by the Euler equations

$$\partial_t \boldsymbol{V} + \partial_x \boldsymbol{f}(\boldsymbol{V}) = \boldsymbol{0} \tag{8.38}$$

with the vector of conserved quantities

$$\boldsymbol{V} = \begin{pmatrix} V_1 \\ V_2 \\ V_3 \end{pmatrix} = \begin{pmatrix} \varrho \\ \varrho u \\ E \end{pmatrix} \tag{8.39}$$

and with the flux

$$\boldsymbol{f}(\boldsymbol{V}) = \begin{pmatrix} \varrho u \\ \varrho u^2 + p \\ u(E+p) \end{pmatrix}, \tag{8.40}$$

as given after Eq. (1.14) in Section 1.1, where we wrote $q = \varrho u$. The total energy E is connected with the internal energy e by means of the equation

$$E = \varrho \left(\frac{1}{2} u^2 + p \right), \tag{8.41}$$

and we employ an equation of state for ideal gases in the form

$$e = e(p, \varrho) = \frac{p}{(\gamma - 1)\varrho}. \tag{8.42}$$

The quantity

$$c = \sqrt{\frac{\gamma p}{\varrho}} \tag{8.43}$$

is the local velocity of sound; cf. (1.66). Although we have already computed the Jacobian $\boldsymbol{Jf}(\boldsymbol{V})$ in Section 1.1, we provide three different forms for easy reference here:

$$\boldsymbol{Jf}(\boldsymbol{V}) = \begin{pmatrix} 0 & 1 & 0 \\ -\frac{1}{2}(\gamma-3)\left(\frac{V_2}{V_1}\right)^2 & (3-\gamma)\left(\frac{V_2}{V_1}\right) & \gamma - 1 \\ -\frac{\gamma V_2 V_3}{V_1^2} + (\gamma-1)\left(\frac{V_2}{V_1}\right)^3 & \frac{\gamma V_3}{V_1} - \frac{3}{2}(\gamma-1)\left(\frac{V_2}{V_1}\right)^2 & \gamma\left(\frac{V_2}{V_1}\right) \end{pmatrix}, \tag{8.44}$$

$$\boldsymbol{Jf}(\boldsymbol{V}) = \begin{pmatrix} 0 & 1 & 0 \\ -\frac{1}{2}(\gamma-3)u^2 & (3-\gamma)u & \gamma - 1 \\ \frac{1}{2}(\gamma-2)u^3 - \frac{c^2 u}{\gamma-1} & \frac{3-2\gamma}{2}u^2 + \frac{c^2}{\gamma-1} & \gamma u \end{pmatrix}, \tag{8.45}$$

and, introducing the *enthalpy*

$$H = \frac{E+p}{\varrho} = \frac{1}{2}u^2 + e + \frac{p}{\varrho} \,,$$

the Jacobian can again be rewritten to yield

$$Jf(V) = \begin{pmatrix} 0 & 1 & 0 \\ -\frac{1}{2}(\gamma-3)u^2 & (3-\gamma)u & \gamma-1 \\ u\left(\frac{1}{2}(\gamma-1)u^2 - H\right) & H-(\gamma-1)u^2 & \gamma u \end{pmatrix} . \tag{8.46}$$

The eigenvalues of $Jf(V)$ can be computed to give (cf. (1.15))

$$\lambda_1 = u - c \,, \quad \lambda_2 = u \,, \quad \lambda_3 = u + c \,, \tag{8.47}$$

and the corresponding eigenvectors are found to be

$$\mathbf{K}^{(1)} = \begin{pmatrix} 1 \\ u-c \\ H-uc \end{pmatrix} , \quad \mathbf{K}^{(2)} = \begin{pmatrix} 1 \\ u \\ \frac{1}{2}u^2 \end{pmatrix} , \quad \mathbf{K}^{(3)} = \begin{pmatrix} 1 \\ u+c \\ H+uc \end{pmatrix} . \tag{8.48}$$

For later reference, we rewrite the Euler equations once more – now in terms of the variables density, velocity, and entropy:

$$s = c_v \ln\left(\frac{p}{\varrho^\gamma}\right) + \text{const} \,,$$

where c_v is the heat capacity at constant volume. In this formulation, we get the following equation for the pressure:

$$p = C\varrho^\gamma \exp\left(\frac{s}{c_v}\right) , \tag{8.49}$$

where C denotes a constant and the vector of unknowns is

$$\mathbf{W} = \begin{pmatrix} \varrho \\ u \\ s \end{pmatrix} ,$$

while the Euler equations are given in quasilinear form (this is also called the *entropy formulation*) as

$$\partial_t \mathbf{W} + \mathbf{A}(\mathbf{W})\partial_x \mathbf{W} = \mathbf{0} \tag{8.50}$$

with the matrix

$$\mathbf{A}(\mathbf{W}) = \begin{pmatrix} u & \varrho & 0 \\ \frac{c^2}{\varrho} & u & \frac{1}{\varrho}\frac{\partial p}{\partial s} \\ 0 & 0 & u \end{pmatrix} . \tag{8.51}$$

The eigenvalues of $\boldsymbol{A}$ are given exactly by (8.47), while the eigenvectors are now

$$\boldsymbol{K}_s^{(1)} = \begin{pmatrix} 1 \\ -\frac{c}{\varrho} \\ 0 \end{pmatrix}, \quad \boldsymbol{K}_s^{(2)} = \begin{pmatrix} -\frac{\partial p}{\partial s} \\ 0 \\ c^2 \end{pmatrix}, \quad \boldsymbol{K}_s^{(3)} = \begin{pmatrix} 1 \\ \frac{c}{\varrho} \\ 0 \end{pmatrix}. \tag{8.52}$$

Each of the eigenvalues λ_i can be interpreted as a characteristic speed, and there is a *characteristic field*, called the *$\boldsymbol{K}^{(i)}$-field* or the i-th characteristic field, associated with each speed. A $\boldsymbol{K}^{(i)}$-field is said to be *linearly degenerated* if, for all states $\boldsymbol{V}$, the equation

$$\left\langle \nabla\lambda_i(\boldsymbol{V}), \boldsymbol{K}^{(i)} \right\rangle = 0$$

holds. According to Section 3.3, a $\boldsymbol{K}^{(i)}$-field is called *genuinely nonlinear* if

$$\left\langle \nabla\lambda_i(\boldsymbol{V}), \boldsymbol{K}^{(i)} \right\rangle \neq 0$$

holds. If a discontinuity occurs in the $\boldsymbol{K}^{(i)}$-field, we shall accept it as an *admissible discontinuity* if the *Rankine–Hugoniot condition*

$$[\boldsymbol{f}] = S_i \cdot [\boldsymbol{V}]$$

holds where $[\boldsymbol{V}] = \boldsymbol{V}_\mathrm{R} - \boldsymbol{V}_\mathrm{L}$ and $[\boldsymbol{f}] = \boldsymbol{f}_\mathrm{R} - \boldsymbol{f}_\mathrm{L} = \boldsymbol{f}(\boldsymbol{V}_\mathrm{R}) - \boldsymbol{f}(\boldsymbol{V}_\mathrm{L})$ denote the jumps across the discontinuity with values $\boldsymbol{V}_\mathrm{R}$ and $\boldsymbol{V}_\mathrm{L}$ to the right and the left, respectively.[143) S_i is the speed of the discontinuity.

An important concept which we will exploit shortly is that of (generalized) *Riemann invariants*; see Section 3.3. Given an arbitrary one-dimensional quasilinear system

$$\partial_t \boldsymbol{w} + \boldsymbol{A}(\boldsymbol{w})\partial_x \boldsymbol{w} = \boldsymbol{0}$$

where $\boldsymbol{w} = (w_1, w_2, w_3)$, we consider the $\boldsymbol{K}^{(i)}$-field corresponding to the eigenvalue λ_i and eigenvector

$$\boldsymbol{K}^{(i)} = \left(K_1^{(i)}, K_2^{(i)}, K_3^{(i)}\right).$$

(Generalized) Riemann invariants are then solutions of the system

$$\frac{\mathrm{d}w_1}{K_1^{(i)}} = \frac{\mathrm{d}w_2}{K_2^{(i)}} = \frac{\mathrm{d}w_3}{K_3^{(i)}}. \tag{8.53}$$

The Riemann invariants are indeed invariant across waves!

143) See Section 2.3

8.7 Elementary Waves

In a Riemann problem

$$\partial_t V + \partial_x f(V) = 0$$

$$V(x,0) = V_0(x) = \begin{cases} V_L\,; & x < 0 \\ V_R\,; & x > 0 \end{cases}$$

where $V = (V_1, V_2, V_3)$, we are now looking for *similarity solutions*, i.e., solutions of the form $V\left(\frac{x}{t}\right)$. These solutions are given by four constant states which are seperated by three waves, as shown in Fig. 8.4.

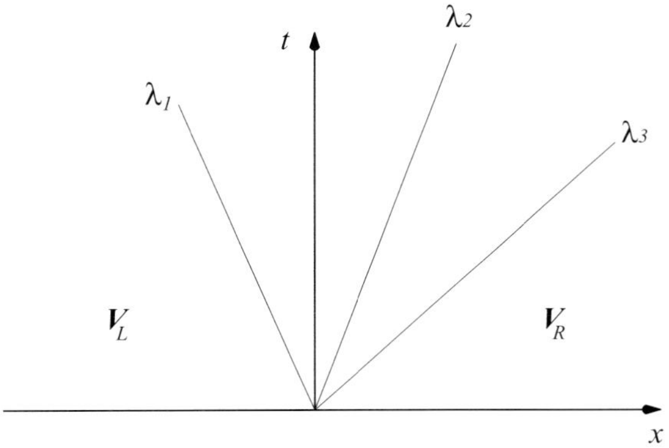

Fig. 8.4 A similarity solution.

In gas dynamics, we know of three different types of waves (see also Section 2.3):

1. *Shocks:* In a genuinely nonlinear field, two constant states V_L and V_R are connected by an admissible jump discontinuity that additionally satisfies the entropy condition, i.e., the Rankine–Hugoniot condition

   $$f(V_R) - f(V_L) = S_i(V_R - V_L)$$

   as well as

   $$\lambda_i(V_L) > S_i > \lambda_i(V_R)\,,$$

   both hold; see Section 3.3.

2. *Contact discontinuities:* In a linearly degenerated field, two constant states V_L and V_R are connected by a jump discontinuity if the Rankine–Hugoniot condition

   $$f(V_R) - f(V_L) = S_i(V_R - V_L)$$

holds, the Riemann invariants[144] are constant:

$$\frac{\mathrm{d}w_1}{K_1^{(i)}} = \frac{\mathrm{d}w_2}{K_2^{(i)}} = \frac{\mathrm{d}w_3}{K_3^{(i)}} ,$$

and the characteristics are parallel:

$$\lambda_i(\mathbf{V}_\mathrm{L}) = \lambda_i(\mathbf{V}_\mathrm{R}) = S_i .$$

3. *Rarefaction waves:* In a genuinely nonlinear field, two constant states $\mathbf{V}_\mathrm{L}$ and $\mathbf{V}_\mathrm{R}$ are connected by a rarefaction wave if the Riemann invariants are constant:

$$\frac{\mathrm{d}w_1}{K_1^{(i)}} = \frac{\mathrm{d}w_2}{K_2^{(i)}} = \frac{\mathrm{d}w_3}{K_3^{(i)}} ,$$

and the characteristics diverge:

$$\lambda_i(\mathbf{V}_\mathrm{L}) < \lambda_i(\mathbf{V}_\mathrm{R}) .$$

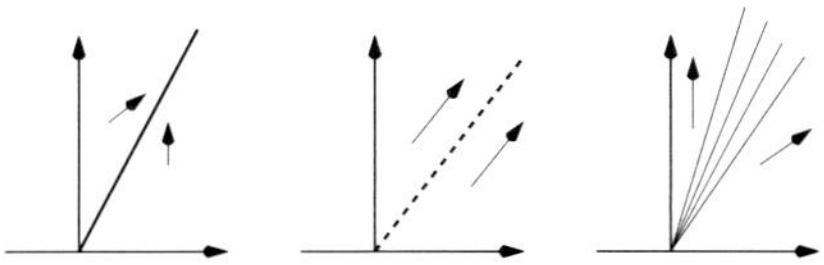

Fig. 8.5 Shock, contact discontinuity, and rarefaction wave (*left to right*).

Returning to the case of the Euler equations, we look at the eigenvalues (8.47) and the corresponding eigenvectors (8.48), and compute for the $\boldsymbol{K}^{(2)}$-field

$$\nabla\lambda_2(\mathbf{V}) = \left(\frac{\partial\lambda_2}{\partial V_1}, \frac{\partial\lambda_2}{\partial V_2}, \frac{\partial\lambda_2}{\partial V_3}\right) = \left(-\frac{u}{\varrho}, \frac{1}{\varrho}, 0\right)$$

so that

$$\left\langle \nabla\lambda_2(\mathbf{V}), \boldsymbol{K}^{(2)} \right\rangle = 0 .$$

Hence, the $\boldsymbol{K}^{(2)}$-field is linearly degenerate. Analogously, it can easily been shown that the $\boldsymbol{K}^{(1)}$- and the $\boldsymbol{K}^{(3)}$-fields are genuinely nonlinear.

Hence, the middle wave corresponding to the wave speed u is always a contact discontinuity. Whether the left and right waves are shock or rarefaction waves depends on the initial values of the Riemann problem.

If we consider the Riemann invariants in the conservative variables $\mathbf{V}$, we get

$$\frac{\mathrm{d}\varrho}{1} = \frac{\mathrm{d}(\varrho u)}{u} = \frac{\mathrm{d}E}{\frac{1}{2}u^2} ,$$

from which we conclude $\mathrm{d}E = \frac{1}{2}u^2\,\mathrm{d}\varrho$, and hence $E = \frac{1}{2}u^2\varrho + \mathrm{const}$. Since we know that $E = \varrho(\frac{1}{2}u^2 + e)$, it follows that the constant is given by ϱe, i.e.,

$$\varrho e = \mathrm{const}$$

144) We write w_k because these may be different variables from the V_k.

and since $e = p/((\gamma - 1)\varrho)$ for an ideal gas, we conclude

$$\text{const} = \varrho e = \frac{p}{\gamma - 1} \quad \Rightarrow \quad p = \text{const}\ .$$

Additionally, we see that $\mathrm{d}(\varrho u) = u\ \mathrm{d}\varrho$, so that

$$u = \text{const}\ .$$

We thus conclude that *pressure and velocity stay unchanged across a contact discontinuity, while the density may jump.*

We now consider rarefaction waves. Employing the eigenvectors (8.52) of the Euler equations in entropy formulation (8.50), the Riemann invariants for $\lambda_1 = u - c$ give

$$\frac{\mathrm{d}\varrho}{1} = \frac{\mathrm{d}u}{-\frac{c}{\varrho}} = \frac{\mathrm{d}s}{0}\ ,$$

from which we conclude

$$u + \int \frac{c}{\varrho}\ \mathrm{d}\varrho = \text{const}\ , \qquad s = \text{const}\ . \tag{8.54}$$

For $\lambda_3 = u + c$, we analogously obtain

$$u - \int \frac{c}{\varrho}\ \mathrm{d}\varrho = \text{const}\ , \qquad s = \text{const}\ . \tag{8.55}$$

For the pressure in the entropy formulation, we previously (see (8.49)) found that

$$p = C\varrho^{\gamma} \exp\left(\frac{s}{c_v}\right) .$$

We can now compute

$$\begin{aligned} \partial_t s &= c_v \frac{\varrho^{\gamma}}{p} \partial_t \left(\frac{p}{\varrho^{\gamma}}\right) = c_v \frac{1}{p}\left(\partial_t p - \frac{\gamma p}{\varrho}\partial_t \varrho\right) \\ &= \frac{c_v}{p}\left(\partial_t p - c^2 \partial_t \varrho\right) , \end{aligned} \tag{8.56}$$

where the last equation follows from the definition of the velocity of sound. Analogously, we find that

$$\partial_x s = \frac{c_v}{p}\left(\partial_x p - c^2 \partial_x \varrho\right) . \tag{8.57}$$

Using the first equation in the system of Euler equations in conservative variables, in the case of smooth flow we find that

$$\partial_t \varrho + u \partial_x \varrho + \varrho \partial_x u = 0\ , \tag{8.58}$$

and using the second equation along with (8.58), we get

$$\partial_t u + u\partial_x u + \frac{1}{\varrho}\partial_x p = 0 \,. \tag{8.59}$$

Using (8.58) and (8.59), the third equation in the Euler system (energy equation) yields

$$\partial_t p + \varrho c^2 \partial_x u + u\partial_x p = 0 \,, \tag{8.60}$$

hence

$$\partial_t p = -\varrho c^2 \partial_x u - u\partial_x p \,.$$

Inserting this into (8.56) and exploiting (8.57) yields

$$\partial_t s + u\partial_x s = 0 \,. \tag{8.61}$$

In rarefaction waves, the entropy satisfies a linear transport equation; in other words, s = const on the curves $\frac{\mathrm{d}x}{\mathrm{d}t} = u$. With this result in hand, we can conclude from (8.49) that

$$p = C\varrho^{\gamma}$$

for this kind of flow, which is called isentropic (cf. Section 3.1). Hence, $c = \sqrt{\frac{\gamma p}{\varrho}} = \sqrt{\gamma C \varrho^{\gamma-1}}$ is a function of ϱ only, and the integrals in (8.54) and (8.55) can be solved to give the *Riemann invariants across $\boldsymbol{K}^{(1)}$- and $\boldsymbol{K}^{(3)}$-fields*:

$$\left.\begin{aligned} I_{\mathrm{L}}(u,c) = u + \tfrac{2c}{\gamma-1} &= \text{const} \\ s &= \text{const} \end{aligned}\right\} \quad \lambda_1 = u - c \tag{8.62}$$

$$\left.\begin{aligned} I_{\mathrm{R}}(u,c) = u - \tfrac{2c}{\gamma-1} &= \text{const} \\ s &= \text{const.} \end{aligned}\right\} \quad \lambda_3 = u + c \,. \tag{8.63}$$

It remains for us to discuss shocks which are admissible jump discontinuities in the genuinely nonlinear fields belonging to $\boldsymbol{K}^{(1)}$ and $\boldsymbol{K}^{(3)}$, respectively. Consider the $\boldsymbol{K}^{(3)}$-field and a right-facing shock with shock velocity S_3 (see Fig. 8.6).

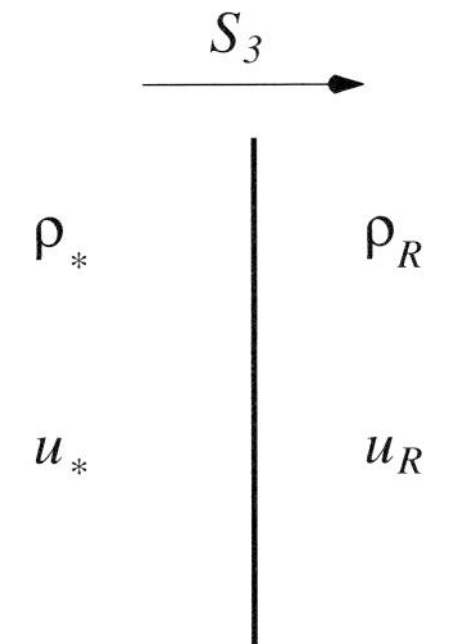

Fig. 8.6 Moving shock in the $\boldsymbol{K}^{(3)}$-field.

The right state $\boldsymbol{V}_{\mathrm{R}}$ is known in the Riemann problem, while the state to the left of the $\boldsymbol{K}^{(3)}$-wave is unknown. In order to eliminate the shock speed, we transform the problem into a steady shock. The transformation will not change density and pressure, but rather velocity:

$$\begin{aligned}\hat{u}_* &= u_* - S_3\ ,\\ \hat{u}_{\mathrm{R}} &= u_{\mathrm{R}} - S_3\ ,\end{aligned}$$

see Fig. 8.7.

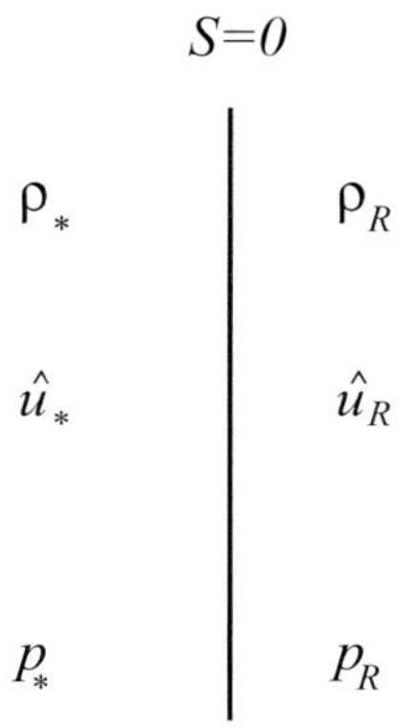

Fig. 8.7 Steady shock.

The Rankine–Hugoniot condition for shock velocity 0 is

$$\varrho_* \hat{u}_* = \varrho_{\mathrm{R}} \hat{u}_{\mathrm{R}} \tag{8.64}$$

$$\varrho_* \hat{u}_*^2 + p_* = \varrho_{\mathrm{R}} \hat{u}_{\mathrm{R}}^2 + p_{\mathrm{R}} \tag{8.65}$$

$$\hat{u}_*(\hat{E}_* + p_*) = \hat{u}_{\mathrm{R}}(\hat{E}_{\mathrm{R}} + p_*)\ . \tag{8.66}$$

With the help of (8.41), the left hand side of (8.66) can be written as

$$\hat{u}_* \varrho_* \left(\frac{1}{2}\hat{u}_*^2 + \left(e_* + \frac{p_*}{\varrho_*}\right)\right),$$

while the right hand side is

$$\hat{u}_{\mathrm{R}} \varrho_{\mathrm{R}} \left(\frac{1}{2}\hat{u}_{\mathrm{R}}^2 + \left(e_{\mathrm{R}} + \frac{p_{\mathrm{R}}}{\varrho_{\mathrm{R}}}\right)\right).$$

Employing the *specific enthalpy* $h = e + \frac{p}{\varrho}$ yields

$$h_* = e_* + \frac{p_*}{\varrho_*}\ , \qquad h_{\mathrm{R}} = e_{\mathrm{R}} + \frac{p_{\mathrm{R}}}{\varrho_{\mathrm{R}}}\ . \tag{8.67}$$

With (8.64) and (8.66), we get

$$\frac{1}{2}\hat{u}_*^2 + h_* = \frac{1}{2}\hat{u}_{\mathrm{R}}^2 + h_{\mathrm{R}}\ . \tag{8.68}$$

Inserting (8.64) into (8.65) yields

$$\varrho_* \hat{u}_*^2 = (\varrho_R \hat{u}_R)\hat{u}_R + p_R - p_* = (\varrho_* \hat{u}_*)\frac{\varrho_* \hat{u}_*}{\varrho_R} + p_R - p_* \,,$$

from which we derive

$$\hat{u}_*^2 = \left(\frac{\varrho_R}{\varrho_*}\right)\left[\frac{p_R - p_*}{\varrho_R - \varrho_*}\right] . \tag{8.69}$$

Analogously, we get

$$\hat{u}_R^2 = \left(\frac{\varrho_*}{\varrho_R}\right)\left[\frac{p_R - p_*}{\varrho_R - \varrho_*}\right] . \tag{8.70}$$

Inserting (8.69) and (8.70) into (8.68) yields

$$h_* - h_R = \frac{1}{2}(p_* - p_R)\left[\frac{\varrho_* + \varrho_R}{\varrho_* \varrho_R}\right] . \tag{8.71}$$

Now suppose that the internal energy e is given by (8.42). Then it follows from (8.71) with the help of (8.67) that

$$e_* - e_R = \frac{1}{2}(p_* + p_R)\left[\frac{\varrho_* - \varrho_R}{\varrho_* \varrho_R}\right] , \tag{8.72}$$

and with (8.42)

$$\frac{\varrho_*}{\varrho_R} = \frac{\frac{p_*}{p_R} + \frac{\gamma-1}{\gamma+1}}{\frac{\gamma-1}{\gamma+1} \cdot \frac{p_*}{p_R} + 1} . \tag{8.73}$$

If we now introduce the *Mach number* $M = \frac{u}{a}$ as

$$M_R = \frac{u_R}{c_R} \,, \qquad M_S = \frac{S_3}{c_R} \,,$$

where M_S is called the *shock Mach number*, then it follows from (8.69) and (8.73) that

$$\frac{\varrho_*}{\varrho_R} = \frac{(\gamma+1)(M_R - M_S)^2}{(\gamma-1)(M_R - M_S)^2 + 2} \tag{8.74}$$

$$\frac{p_*}{p_R} = \frac{2\gamma(M_R - M_S)^2 - (\gamma-1)}{\gamma+1} . \tag{8.75}$$

From (8.75), we see

$$M_R - M_S = -\sqrt{\frac{\gamma+1}{2\gamma} \cdot \frac{p_*}{p_R} + \frac{\gamma-1}{2\gamma}} \,, \tag{8.76}$$

and this gives the following expression for the shock speed:

$$S_3 = u_R + c_R\sqrt{\frac{\gamma+1}{2\gamma} \cdot \frac{p_*}{p_R} + \frac{\gamma-1}{2\gamma}} . \tag{8.77}$$

Note that in the case $\frac{p_*}{p_R} \longrightarrow 1$, we get $S_3 \longrightarrow \lambda_3 = u_R + c_R$, as it should be. Finally, we find u_* from (8.64) to be

$$u_* = \left(1 - \frac{\varrho_R}{\varrho_*}\right) S_3 + u_R \frac{\varrho_R}{\varrho_*} . \tag{8.78}$$

What have we achieved so far? *The equations (8.74), (8.75), (8.78) define a state*

$$(\varrho_*, u_*, p_*)$$

behind a shock in the $\boldsymbol{K}^{(3)}$*-field (i.e., behind a right-facing shock) if* M_S *(or* S_3*) is given.*

The computations above are tedious but simple. Nevertheless, in principle we have to go through them again now to get the state to the right of a shock in the $\boldsymbol{K}^{(1)}$ field, i.e., a left-facing shock. Since the computations are completely analogous to the previous ones, we will just state the results here. Defining

$$\hat{u}_L = u_L - S_1 , \quad \hat{u}_* = u_* - S_1$$

and the Mach numbers

$$M_L = \frac{u_L}{c_L} , \quad M_S = \frac{S_1}{c_L} ,$$

one arrives at

$$\frac{\varrho_*}{\varrho_L} = \frac{\frac{p_*}{p_L} + \frac{\gamma-1}{\gamma+1}}{\frac{\gamma-1}{\gamma+1} \cdot \frac{p_*}{p_L} + 1} , \tag{8.79}$$

and

$$\frac{\varrho_*}{\varrho_L} = \frac{(\gamma + 1)(M_L - M_S)^2}{(\gamma - 1)(M_L - M_S)^2 + 2} . \tag{8.80}$$

Furthermore,

$$\frac{p_*}{p_L} = \frac{2\gamma(M_L - M_S)^2 - (\gamma - 1)}{\gamma + 1} . \tag{8.81}$$

The shock speed S_1 follows from either (8.80) or (8.81). Starting from the latter yields

$$M_L - M_S = \sqrt{\frac{\gamma + 1}{2\gamma} \cdot \frac{p_*}{p_L} + \frac{\gamma - 1}{2\gamma}}$$

and hence

$$S_1 = u_L - c_L \sqrt{\frac{\gamma + 1}{2\gamma} \cdot \frac{p_*}{p_L} + \frac{\gamma - 1}{2\gamma}} . \tag{8.82}$$

The velocity u_* is given by

$$u_* = \left(1 - \frac{\varrho_L}{\varrho_*}\right) S_1 + u_L \frac{\varrho_L}{\varrho_*} . \tag{8.83}$$

We have now described the complete theory of gas dynamics within a Riemann problem in detail.

8.8 The Complete Solution to the Riemann Problem

We can now obtain the solution to the Riemann problem, which is at the heart of Godunov's method. The required gas dynamics were all developed above. There are four different cases, which are depicted in Fig. 8.8. It will be convenient to use *primitive variables*, which we denote by

$$w = (\varrho, u, p)\,,$$

and our general problem statement is given visually in Fig. 8.9.

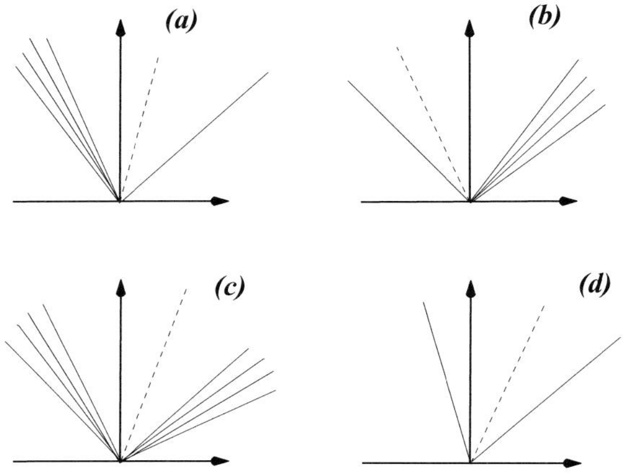

Fig. 8.8 Possible wave configurations.

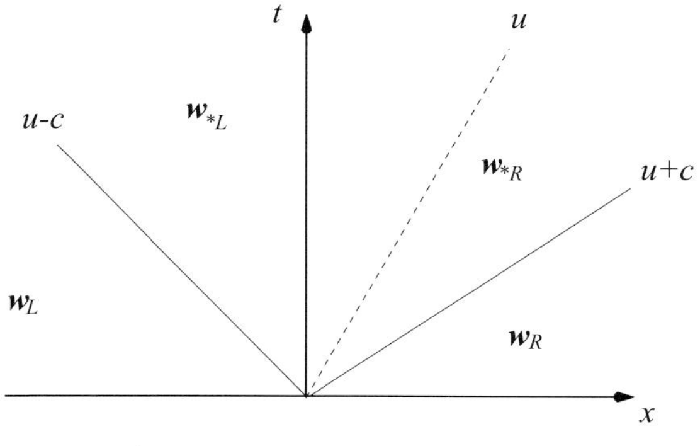

Fig. 8.9 Problem statement.

The left and right states w_L and w_R, respectively, are given. Then there are two unknown states, denoted by w_{*L} (between the left wave and the contact discontinuity) and w_{*R} (between the contact discontinuity and the right wave). The left and right waves may be rarefaction waves or shocks.

We know already that

$$p_{*L} = p_{*R} =: p_* \tag{8.84}$$

and

$$u_{*L} = u_{*R} =: u_* \ . \tag{8.85}$$

The Riemann problem for the one-dimensional Euler equations is solved if we can compute the values

$$p_*, u_*, \varrho_{*L}, \varrho_{*R} \ .$$

The solution rests on the following result:[145)]

Theorem 8.8

Computing the pressure and velocity: The pressure p_* is the solution to the algebraic equation

$$f(p, \boldsymbol{w}_L, \boldsymbol{w}_R) := f_L(p, \boldsymbol{w}_L) + f_R(p, \boldsymbol{w}_R) + \Delta u \tag{8.86}$$

with

$$\Delta u = u_R - u_L$$

and

$$f_L(p, \boldsymbol{w}_L) = \begin{cases} (p - p_L)\sqrt{\dfrac{A_L}{p + B_L}} & ; \quad p > p_L \quad \text{(shock)} \\ \dfrac{2c_L}{\gamma - 1}\left(\left(\dfrac{p}{p_L}\right)^{\frac{\gamma-1}{2\gamma}} - 1\right) & ; \quad p \le p_L \quad \text{(rarefaction)} \end{cases} , \tag{8.87}$$

$$f_R(p, \boldsymbol{w}_R) = \begin{cases} (p - p_R)\sqrt{\dfrac{A_R}{p + B_R}} & ; \quad p > p_R \quad \text{(shock)} \\ \dfrac{2c_R}{\gamma - 1}\left(\left(\dfrac{p}{p_R}\right)^{\frac{\gamma-1}{2\gamma}} - 1\right) & ; \quad p \le p_R \quad \text{(rarefaction)} \end{cases} , \tag{8.88}$$

$$\left.\begin{array}{ll} A_L = \dfrac{2}{(\gamma + 1)\varrho_L} , & B_L = \frac{\gamma-1}{\gamma+1} p_L \\ A_R = \dfrac{2}{(\gamma + 1)\varrho_R} , & B_R = \frac{\gamma-1}{\gamma+1} p_R \end{array}\right\} , \tag{8.89}$$

145) We closely follow the presentation given in E.F. Toro: *Riemann solvers and numerical methods for fluid dynamics, 2nd ed.* Berlin: Springer 1999.

and the velocity u_* is given by

$$u_* = \frac{1}{2}(u_L + u_R) + \frac{1}{2}\left(f_R(p_*, \boldsymbol{w}_R) - f_L(p_*, \boldsymbol{w}_L)\right) . \tag{8.90}$$

Proof:

The proof proceeds by distinguishing between different possible cases.

(I) f_L *in the case of a left-facing shock with speed* S_L. In this case, we again transform to a steady shock, defining

$$\hat{u}_L = u_L - S_L , \quad \hat{u}_* = u_* - S_L . \tag{8.91}$$

The Rankine–Hugoniot conditions then read as

$$\varrho_L \hat{u}_L = \varrho_{*L} \hat{u}_* \tag{8.92}$$

$$\varrho_L \hat{u}_L^2 + p_L = \varrho_{*L} \hat{u}_*^2 + p_* \tag{8.93}$$

$$\hat{u}_L(\hat{E}_L + p_L) = \hat{u}_*(\hat{E}_{*L} + p_*) . \tag{8.94}$$

Now we define the *mass flux*

$$Q_L = \varrho_L \hat{u}_L = \varrho_{*L} \hat{u}_* . \tag{8.95}$$

Rewriting (8.93) as

$$(\varrho_L \hat{u}_L)\hat{u}_L + p_L = (\varrho_{*L} \hat{u}_*)\hat{u}_* + p_*$$

and using (8.95), we arrive at

$$Q_L = -\frac{p_* - p_L}{\hat{u}_* - \hat{u}_L} , \tag{8.96}$$

but from (8.91) it is seen that $\hat{u}_L - \hat{u}_* = u_L - u_*$; based on this and (8.96), we conclude that

$$Q_L = -\frac{p_* - p_L}{u_* - u_L} , \tag{8.97}$$

and therefore

$$u_* = u_L - \frac{p_* - p_L}{Q_L} . \tag{8.98}$$

Equation (8.95) yields $\hat{u}_L = \frac{Q_L}{\varrho_L}$, $\hat{u}_* = \frac{Q_L}{\varrho_{*L}}$, which we insert into (8.96) to arrive at

$$Q_L^2 = -\frac{p_* - p_L}{\frac{1}{\varrho_{*L}} - \frac{1}{\varrho_L}} . \tag{8.99}$$

Following (8.79), we know that

$$\varrho_{*\mathrm{L}} = \varrho_{\mathrm{L}} \left(\frac{\frac{\gamma-1}{\gamma+1} + \frac{p_*}{p_{\mathrm{L}}}}{\frac{\gamma-1}{\gamma+1}\frac{p_*}{p_{\mathrm{L}}} + 1} \right) , \tag{8.100}$$

and inserting this into (8.99) gives

$$Q_{\mathrm{L}} = \sqrt{\frac{p_* + B_{\mathrm{L}}}{A_{\mathrm{L}}}} , \tag{8.101}$$

which in turn, and using (8.98), results in

$$u_* = u_{\mathrm{L}} - f_{\mathrm{L}}(p_*, \boldsymbol{w}_{\mathrm{L}}) . \tag{8.102}$$

(II) f_{L} *in the case of a left-facing rarefaction wave.* In this case, the flow is isentropic, and hence

$$p = C\varrho^{\gamma} . \tag{8.103}$$

The constant C is determined at the left state, i.e., $p_{\mathrm{L}} = C\varrho_{\mathrm{L}}^{\gamma}$, and hence $C = \frac{p_{\mathrm{L}}}{\varrho_{\mathrm{L}}^{\gamma}}$. Then it follows from (8.103) that

$$\varrho_{*\mathrm{L}} = \varrho_{\mathrm{L}} \sqrt[\gamma]{\frac{p_*}{p_{\mathrm{L}}}} . \tag{8.104}$$

We know that the Riemann invariant $I_{\mathrm{L}}(u, c)$ is constant across a rarefaction wave:

$$u_{\mathrm{L}} + \frac{2c_{\mathrm{L}}}{\gamma - 1} = u_* + \frac{2c_{*\mathrm{L}}}{\gamma - 1} . \tag{8.105}$$

The speed of sound is given by definition as $c_{*\mathrm{L}} = \sqrt{\frac{\gamma p_*}{\varrho_{*\mathrm{L}}}}$, and so, using (8.104), we have

$$c_{*\mathrm{L}} = \sqrt{\frac{\gamma p_*}{\varrho_{\mathrm{L}} \left(\frac{p_*}{p_{\mathrm{L}}}\right)^{\frac{1}{\gamma}}}} = c_{\mathrm{L}} \left(\frac{p_*}{p_{\mathrm{L}}}\right)^{\frac{\gamma-1}{2\gamma}} , \tag{8.106}$$

and (8.105) then leads to

$$u_* = u_{\mathrm{L}} - f_{\mathrm{L}}(p_*, \boldsymbol{w}_{\mathrm{L}}) . \tag{8.107}$$

Now, turning our attention to right-facing waves, we proceed more rapidly because everything is pretty much analogous to the left-facing cases.

(III) *f_R in the case of a right-facing shock with speed S_R*. Here, we again transform to a steady shock via

$$\hat{u}_R - S_R\,, \qquad \hat{u}_* = u_* - S_R$$

and write down the Rankine–Hugoniot conditions as

$$\begin{aligned} \varrho_{*R}\hat{u}_* &= \varrho_R\hat{u}_R \\ \varrho_{*R}\hat{u}_*^2 + p_* &= \varrho_R\hat{u}_R^2 + p_R \\ \hat{u}_*(\hat{E}_{*R} + p_*) &= \hat{u}_R(\hat{E}_R + p_R)\,. \end{aligned}$$

The mass flux is now defined as

$$Q_R = -\varrho_{*R}\hat{u}_* = -\varrho_R\hat{u}_R\,.$$

In complete analogy to what we did in the case of a left-facing shock, we arrive at

$$Q_R = \sqrt{\frac{p_* + B_R}{A_R}}$$

and

$$u_* = u_R + f_R(p_*, \boldsymbol{w}_R)\,. \tag{8.108}$$

(IV) *f_R in the case of a right-facing rarefaction wave.* This case is also completely analogous to left-facing rarefaction. We finally arrive at the speed of sound,

$$c_{*R} = c_R\left(\frac{p_*}{p_R}\right)^{\frac{\gamma-1}{2\gamma}}\,,$$

and at the velocity,

$$u_* = u_R + f_R(p_*, \boldsymbol{w}_R)\,. \tag{8.109}$$

In order to conclude, let us state that by eliminating u_* from (8.98) or (8.107) and (8.108) or (8.109), we get $f_L(p_*, \boldsymbol{w}_L) + f_R(p_*, \boldsymbol{w}_R) + \Delta u = 0$. Solving this equation for p_* gives u_* depending on the particular case according to (8.98), (8.107), (8.108) or (8.109). One can put all of the cases together in the formula

$$u_* = \frac{1}{2}(u_L + u_R) + \frac{1}{2}\left(f_R(p_*, \boldsymbol{w}_R) - f_L(p_*, \boldsymbol{w}_L)\right)\,.$$

Solving (8.86) for the pressure p_* amounts to solving a nonlinear equation, and iterative methods are required. In the literature, one can find useful hints about the type of methods that are preferentially used.[146] We assume that we now have p_* as well as u_*. It now remains to compute ϱ_{*L} and ϱ_{*R}. This can be done according to the particular flow case.

1. A *left-facing shock* is identified by $p_* > p_L$. After having computed p_* and u_*, we compute ϱ_{*L} with the help of (8.100) and the shock speed from (8.82).

2. In contrast, a *left-facing rarefaction wave* is characterized by $p_* \leq p$. After having computed p_* and u_*, we compute ϱ_{*L} from the isentropic equation

   $$\varrho_{*L} = \varrho_L \left(\frac{p_*}{p_L}\right)^{\frac{1}{\gamma}} ,$$

   and the speed of sound from

   $$c_{*L} = c_L \left(\frac{p_*}{p_L}\right)^{\frac{\gamma-1}{2\gamma}} .$$

   Coming from the left, the beginning of the rarefaction wave (which we call the *head*) has a speed of

   $$S_{\text{head,L}} = u_L - c_L .$$

   The other end of the rarefaction wave (on the right of it) is the *tail* of the wave, with speed

   $$S_{\text{tail,L}} = u_* - c_{*L} .$$

   If we denote the state at any point *within* the rarefaction wave by $\boldsymbol{w}_* = (\varrho, u, p)$, and remember the characteristic equation

   $$\frac{\mathrm{d}x}{\mathrm{d}t} = \frac{x}{t} = u - a$$

   as well as the constancy of the Riemann invariant

   $$I_L(u, c) = \text{const} \quad \Rightarrow \quad u_L + \frac{2c_L}{\gamma - 1} = u + \frac{2c}{\gamma - 1} ,$$

   then we can solve both equations simultaneously for u and c. Together with the definition of the speed of sound as well as the isentropic equation, we get

   $$\boldsymbol{w}_* = \begin{cases} \varrho = \varrho_L \left(\frac{2}{\gamma+1} + \frac{\gamma-1}{(\gamma+1)c_L}\left(u_L - \frac{x}{t}\right)\right)^{\frac{2}{\gamma-1}} \\ u = \frac{2}{\gamma+1}\left(c_L + \frac{\gamma-1}{2}u_L + \frac{x}{t}\right) \\ p = p_L \left(\frac{2}{\gamma+1} + \frac{\gamma-1}{(\gamma+1)c_L}\left(u_L - \frac{x}{t}\right)\right)^{\frac{2\gamma}{\gamma-1}} \end{cases} .$$

146) See the very instructive book by Toro cited previously

3. A *right-facing shock* is characterized by $p_* > p_R$. After having computed u_* and p_*, we get

$$\varrho_{*R} = \varrho_R \left(\frac{\frac{p_*}{p_R} + \frac{\gamma-1}{\gamma+1}}{\frac{\gamma-1}{\gamma+1}\frac{p_*}{p_R} + 1} \right)$$

and

$$S_R = u_R + c_R \sqrt{\frac{\gamma+1}{2\gamma}\frac{p_*}{p_R} + \frac{\gamma-1}{2\gamma}} \; .$$

4. A *right-facing rarefaction wave* is identified by $p_* \le p_R$. After computing p_* and u_*, we compute

$$\varrho_{*R} = \varrho_R \left(\frac{p_*}{p_R} \right)^{\frac{1}{\gamma}}$$

and

$$c_{*R} = c_R \left(\frac{p_*}{p_R} \right)^{\frac{\gamma-1}{2\gamma}} \; .$$

The head of the rarefaction wave is now on the right side, with speed

$$S_{\text{head},R} = u_R + c_R \; ,$$

and the tail correspondingly has a speed of

$$S_{\text{tail},R} = u_* + c_{*R} \; .$$

As in the case of the left-facing rarefaction wave, we can compute the state $\boldsymbol{w}_* = (\varrho, u, p)$ anywhere within the wave as

$$\boldsymbol{w}_* = \begin{cases} \varrho = \varrho_R \left(\frac{2}{\gamma+1} - \frac{\gamma-1}{(\gamma+1)c_R}\left(u_R - \frac{x}{t}\right)\right)^{\frac{2}{\gamma-1}} \\ u = \frac{2}{\gamma+1}\left(-c_R + \frac{\gamma-1}{2} u_R + \frac{x}{t}\right) \\ p = p_R \left(\frac{2}{\gamma+1} - \frac{\gamma-1}{(\gamma+1)c_R}\left(u_R - \frac{x}{t}\right)\right)^{\frac{2\gamma}{\gamma-1}} \end{cases} \; .$$

We have given the complete solution to the Riemann problem in gas dynamics, and we are now able to compute the solution – at least iteratively up to any desired accuracy – at any point in space-time.

8.9 The Godunov Scheme in Gas Dynamics

In order to solve the Riemann problem

$$\partial_t \boldsymbol{V} + \partial_x \boldsymbol{f}(\boldsymbol{V}) = \boldsymbol{0} \tag{8.110}$$

$$\boldsymbol{V}_0(x) = \boldsymbol{V}(x,0) = \begin{cases} \boldsymbol{V}_L \; ; & x < 0 \\ \boldsymbol{V}_R \; ; & x > 0 \end{cases} \tag{8.111}$$

for the one-dimensional Euler equations, i.e., for

$$V = \begin{pmatrix} \varrho \\ \varrho u \\ E \end{pmatrix}, \quad f(V) = \begin{pmatrix} \varrho u \\ \varrho u^2 + p \\ u(E+p) \end{pmatrix}$$

where $E = \varrho(u^2/2 + p)$ and with the ideal gas law $e = p/((\gamma - 1)\varrho)$, we could now use the Godunov recipe as described in (8.17) – (8.19). If we denote a numerical approximation to V by $\tilde{V}$, we could then write

$$\tilde{V}_j^{n+1} = \frac{1}{\Delta x}\left[\int_0^{\frac{\Delta x}{2}} V^{\mathrm{R}}(z, t_{n+1}; \tilde{V}_{j-1}^n, \tilde{V}_j^n)\,\mathrm{d}z + \int_{-\frac{\Delta x}{2}}^{0} V^{\mathrm{R}}(z, t_{n+1}; \tilde{V}_j^n, \tilde{V}_{j+1}^n)\,\mathrm{d}z\right], \tag{8.112}$$

where V^{R} denotes the solution to the Riemann problem in conservative variables. However, while (8.112) is a valid Godunov method, it has a particular disadvantage. We know from the last sections how to solve the Riemann problem completely at any point in space-time that we choose. This is a *pointwise* solution, but we would have to integrate according to (8.112). Although there is no conceptional problem in choosing a complicated (or simple) quadrature rule, this approach is seldomly used in practice. The *Godunov method* is in fact based on the following finite difference form:

$$\tilde{V}_j^{n+1} = \tilde{V}_j^n + \frac{\Delta t}{\Delta x}\left(g_{j+\frac{1}{2}}^n - g_{j-\frac{1}{2}}^n\right) \tag{8.113}$$

with numerical flux

$$g_{j+\frac{1}{2}}^n = f\left(V^{\mathrm{R}}(0, t_{n+1}; V_j^n, V_{j+1}^n)\right), \tag{8.114}$$

and so the exact solution to local Riemann problems is only required at the points $x_{j+\frac{1}{2}}$. Another advantage of this implementation comes from the fact that we do not have to work with half the CFL number to avoid wave interactions, as in (8.112).

Why have we invested so much effort in the solution of the Riemann problem in order to now arrive at a numerical method?

There are at least two important reasons. Although other finite difference methods are currently in fashion, the Godunov scheme is still important to the development of modern schemes. It can be seen as the grandfather to a whole family of new schemes called *Godunov-type methods* in which local Riemann problems are solved *approximately*. The *Roe scheme* (named after Phil Roe) is a famous scheme of this kind. A book like ours is not the place to develop these classes of schemes further, but the idea of considering the waves in local Riemann problems is relevant here; it is of great importance in numerical methods and will be so even in the future. The second reason for the effort we invested lies in an important pedagogical fact. Anyone starting out in the fields of modeling in fluid dynamics or numerical modeling should work through the Riemann problem, since this will aid their understanding of trickier situations. There are now attempts to model complicated flows, like multiphase flows, non-Newtonian fluids, flows in the presence of vacuum, low Mach number compressible flows, and so forth. In order to understand not only the phenomenona but also the numerical mathematics required to achieve this, a thorough understanding of the classical Riemann problem is indispensable.

We close with some words on the use of the Godunov method. As it stands, the Godunov method (like the Lax–Friedrichs method) is accurate to the first order, which is not accurate enough for today's requirements. Therefore, recovery algorithms (which we shall describe in some detail for finite volume methods on unstructured grids in Chapter 10) have been developed over the years to increase accuracy. In many of those modern schemes, the Godunov method is still a main building block.

9
Discrete Models on Curvilinear Grids

9.1
Mappings

Although there are commercial finite difference codes on Cartesian meshes where one has to employ sophisticated rules for boundary treatment, the finite difference methods are usually applied on curvilinear body-fitted grids. Since a wide variety of textbooks and research monographs are available on the subject, we will not go much into the details of the generation of curvilinear grids itself here but instead confine ourselves to modeling aspects.

Consider two-dimensional polar coordinates given by

$$\begin{aligned} x^1\left(\xi^1, \xi^2\right) &= \xi^1 \cos\left(\xi^2\right) \\ x^2\left(\xi^1, \xi^2\right) &= \xi^1 \sin\left(\xi^2\right). \end{aligned}$$

Here we have written ξ^1 for the radius and ξ^2 for the angle. Consider a circular obstacle with radius R_1 in Cartesian (x^1, x^2)-space, as shown in Fig. 9.1. If we want to describe some sort of flow around this obstacle, we need to restrict the far field by means of an outer circle, which we assume to have a radius R_2. Using the above transformation, we are now able to address every point between the surface of the obstacle and the farfield circle by prescribing $\xi^1 \in [R_1, R_2]$, $\xi^2 \in [0, 2\pi]$.

How can we interpret the transformation between so-called *physical space* (spanned by the Cartesian coordinates) and the *parameter space* spanned by ξ^1, ξ^2? Suppose that a slit is inserted along the line AB in Fig. 9.1 and then the space

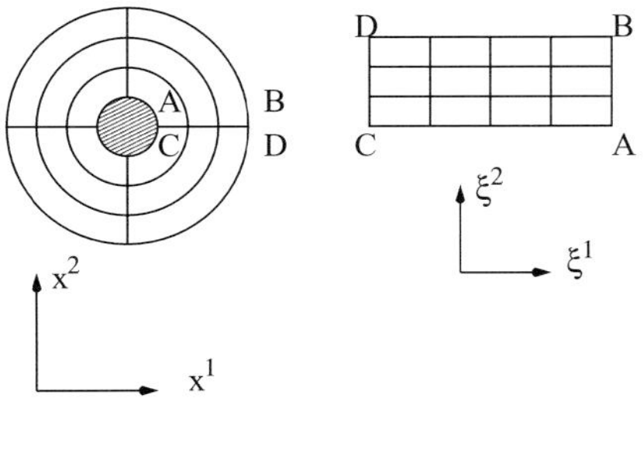

Fig. 9.1 Polar coordinates.

Mathematical Models of Fluid Dynamics. R. Ansorge and T. Sonar

ISBN: 978-3-527-40774-3

between the obstacle and the farfield circle is continuously deformed into a rectangular shape. Note that the points A and C are identical in physical space, as are the points B and D. Due to the slit and the continuous transformation into a rectangular shape, these points now appear to be separated, but they are not. In other words, the boundaries CD and AB are *periodic boundaries*: moving from left to right through the line AB means entering the parameter space again on the left hand side through CD.

The reason for considering the transformation above is easy to see. Finite difference methods are constructed on Cartesian grids, but the grid in physical space is not Cartesian. However, the parameter space is Cartesian and so we can now solve the finite difference equations set up in physical space in the Cartesian parameter space. However, the differential equations describing the flow must be transformed to the parameter domain. Note that we know the transformation relations exactly in the example with the polar coordinates. If we consider an airfoil given as a set of points in the plane (or a set of splines), generally speaking we will not know the transformation exactly.

We now turn our attention to describing the general case in two space dimensions.[147] Three-dimensional transformations are much more demanding due to practical issues, but mathematically there is no difference from the 2D case. In general, a coordinate transformation between a parameter space Π^2 and Cartesian space X^2 should be a *diffeomorphism*, i.e., a smooth, invertible mapping. Employing the canonical basis vectors $\boldsymbol{e}_1, \boldsymbol{e}_2$, we have the Cartesian and parametric coordinates, respectively, given as

$$\boldsymbol{x} = x^1\boldsymbol{e}_1 + x^2\boldsymbol{e}_2 \tag{9.1}$$

$$\boldsymbol{\xi} = \xi^1\boldsymbol{e}_1 + \xi^2\boldsymbol{e}_2 \,. \tag{9.2}$$

Note that we are dealing with the transformations

$$\boldsymbol{x}\colon \begin{cases} \Pi^2 & \longrightarrow X^2 \\ \boldsymbol{\xi} & \longmapsto \boldsymbol{x}(\boldsymbol{\xi}) \end{cases}$$

$$\boldsymbol{\xi}\colon \begin{cases} X^2 & \longrightarrow \Pi^2 \\ \boldsymbol{x} & \longmapsto \boldsymbol{\xi}(\boldsymbol{x}) \end{cases}$$

which are inverse to each other. Writing

$$\boldsymbol{x}\left(\xi^1, \xi^2\right) = \boldsymbol{x}\left(\xi^1(x^1, x^2), \xi^2(x^1, x^2)\right)$$

and differentiating, we get, with the help of (9.1),

$$\frac{\partial \boldsymbol{x}}{\partial x^1} = \boldsymbol{e}_1 = \frac{\partial \boldsymbol{x}}{\partial \xi^1}\frac{\partial \xi^1}{\partial x^1} + \frac{\partial \boldsymbol{x}}{\partial \xi^2}\frac{\partial \xi^2}{\partial x^1}$$

147) A beautiful treatment is given in the book by V.D. Liseikin: *Grid generation methods*. Berlin: Springer 1999.

$$\frac{\partial \boldsymbol{x}}{\partial x^2} = \boldsymbol{e}_2 = \frac{\partial \boldsymbol{x}}{\partial \xi^1}\frac{\partial \xi^1}{\partial x^2} + \frac{\partial \boldsymbol{x}}{\partial \xi^2}\frac{\partial \xi^2}{\partial x^2} .$$

We can write each of the equations in the form of a matrix-vector product as

$$\boldsymbol{e}_i = \boldsymbol{J} \cdot \frac{\partial \boldsymbol{\xi}}{\partial x^i} , \quad i = 1, 2 \tag{9.3}$$

with the help of the Jacobian matrix

$$\boldsymbol{J} := \begin{bmatrix} \frac{\partial x^1}{\partial \xi^1} & \frac{\partial x^1}{\partial \xi^2} \\ \frac{\partial x^2}{\partial \xi^1} & \frac{\partial x^2}{\partial \xi^2} \end{bmatrix} . \tag{9.4}$$

Combining both Eqs. (9.3) into one, we arrive at the matrix equation

$$\boldsymbol{I} = \boldsymbol{J} \cdot \begin{bmatrix} \frac{\partial \xi^1}{\partial x^1} & \frac{\partial \xi^1}{\partial x^2} \\ \frac{\partial \xi^2}{\partial x^1} & \frac{\partial \xi^2}{\partial x^2} \end{bmatrix} , \tag{9.5}$$

where $\boldsymbol{I}$ denotes the two-dimensional unit matrix. We now turn our attention to

$$\boldsymbol{\xi}\left(x^1, x^2\right) = \boldsymbol{\xi}\left(x^1(\xi^1, \xi^2), x^2(\xi^1, \xi^2)\right),$$

which, after differentiation and with the help of (9.2), yields

$$\frac{\partial \boldsymbol{\xi}}{\partial \xi^i} = \boldsymbol{e}_i = \frac{\partial \boldsymbol{\xi}}{\partial x^1}\frac{\partial x^1}{\partial \xi^i} + \frac{\partial \boldsymbol{\xi}}{\partial x^2}\frac{\partial x^2}{\partial \xi^i} ,$$

or, again in the form of a matrix equation,

$$\boldsymbol{I} = \boldsymbol{K} \cdot \begin{bmatrix} \frac{\partial x^1}{\partial \xi^1} & \frac{\partial x^1}{\partial \xi^2} \\ \frac{\partial x^2}{\partial \xi^1} & \frac{\partial x^2}{\partial \xi^2} \end{bmatrix} \tag{9.6}$$

where $\boldsymbol{K}$ is the matrix

$$\boldsymbol{K} = \begin{bmatrix} \frac{\partial \xi^1}{\partial x^1} & \frac{\partial \xi^1}{\partial x^2} \\ \frac{\partial \xi^2}{\partial x^1} & \frac{\partial \xi^2}{\partial x^2} \end{bmatrix} .$$

Comparing this matrix $\boldsymbol{K}$ with (9.5) yields the first fundamental result,

$$\boldsymbol{K} = \boldsymbol{J}^{-1} . \tag{9.7}$$

Hence, if $\det \boldsymbol{J}$ is the Jacobian determinant, then $\det \boldsymbol{K} = 1/\det \boldsymbol{J}$. The elements of the matrices J and $K = J^{-1}$ are related through

$$\frac{\partial \xi^i}{\partial x^j} = (-1)^{i+j}\frac{1}{\det \boldsymbol{J}}\frac{\partial x^{3-j}}{\partial \xi^{3-i}} , \quad i,j = 1, 2$$

$$\frac{\partial x^i}{\partial \xi^j} = (-1)^{i+j}(\det \boldsymbol{J})\frac{\partial \xi^{3-j}}{\partial x^{3-i}} , \quad i,j = 1, 2$$

which can easily be computed from (9.7).

9.2
Transformation Relations

The value $\boldsymbol{x}(\boldsymbol{\xi})$ gives a point in physical space corresponding to a given point $\boldsymbol{\xi}$ in parameter space. Geometrically, each of the vectors $\frac{\partial \boldsymbol{x}}{\partial \xi^i}$ is a *tangent vector* to the grid line ξ^j = const, $j \neq i$. These vectors are called the *covariant basis vectors.* It is easily seen that $\det \boldsymbol{J}$ is the volume of the parallelogram spanned by $\frac{\partial \boldsymbol{x}}{\partial \xi^i}$, $i = 1, 2$. In contrast, the vectors $\nabla \xi^i = \left(\frac{\partial \xi^i}{\partial x^1}, \frac{\partial \xi^i}{\partial x^2}\right)$ are vectors normal to the lines of constant ξ^i. They are called the *contravariant vectors.*

Suppose now that two linearly independent vectors $\boldsymbol{a}_1, \boldsymbol{a}_2 \in \mathbb{R}^2$ are given as well as a further vector $\boldsymbol{b} \in \mathbb{R}^2$ with components b^1, b^2 with respect to the canonical basis $\{\boldsymbol{e}_1, \boldsymbol{e}_2\}$. If we want to represent the vector $\boldsymbol{b}$ as a linear combination of the $\boldsymbol{a}_i$, then we recall from linear algebra that

$$\boldsymbol{b} = \sum_{i=1}^{2} \sum_{j=1}^{2} a^{ij} \left\langle \boldsymbol{b}, \boldsymbol{a}_j \right\rangle \boldsymbol{a}_i \tag{9.8}$$

where a^{ij} are elements of a matrix $\left(a^{ij}\right)$ which is the inverse of the matrix (a_{ij}) defined through

$$a_{ij} = \left\langle \boldsymbol{a}_i, \boldsymbol{a}_j \right\rangle , \quad i, j = 1, 2 .$$

Expanding the vector $\boldsymbol{b}$ in terms of the basis $\left\{\frac{\partial \boldsymbol{x}}{\partial \xi^i}\right\}$, $i = 1, 2$, i.e.,

$$\boldsymbol{b} = \overline{b}^1 \frac{\partial \boldsymbol{x}}{\partial \xi^1} + \overline{b}^2 \frac{\partial \boldsymbol{x}}{\partial \xi^2} ,$$

then the components $\overline{b}^i$ are called the *contravariant components* of $\boldsymbol{b}$. Taking the vectors $\boldsymbol{a}_i$ in (9.8) to be $\boldsymbol{a}_i = \frac{\partial \boldsymbol{x}}{\partial \xi^i}$, $i = 1, 2$, then we obtain

$$\begin{aligned}
\boldsymbol{b} &= \sum_i \sum_j a^{ij} \left\langle \boldsymbol{b}, \boldsymbol{a}_j \right\rangle \boldsymbol{a}_i = \sum_i \sum_j a^{ij} \left\langle \boldsymbol{b}, \frac{\partial \boldsymbol{x}}{\partial \xi^j} \right\rangle \frac{\partial \boldsymbol{x}}{\partial \xi^i} \\
&= \underbrace{\left[a^{11} \left\langle \boldsymbol{b}, \frac{\partial \boldsymbol{x}}{\partial \xi^1} \right\rangle + a^{12} \left\langle \boldsymbol{b}, \frac{\partial \boldsymbol{x}}{\partial \xi^2} \right\rangle \right]}_{=\overline{b}^1} \frac{\partial \boldsymbol{x}}{\partial \xi^1} \\
&\quad + \underbrace{\left[a^{21} \left\langle \boldsymbol{b}, \frac{\partial \boldsymbol{x}}{\partial \xi^1} \right\rangle + a^{22} \left\langle \boldsymbol{b}, \frac{\partial \boldsymbol{x}}{\partial \xi^2} \right\rangle \right]}_{=\overline{b}^2} \frac{\partial \boldsymbol{x}}{\partial \xi^2} ,
\end{aligned}$$

i.e.,

$$\overline{b}^i = \sum_j a^{ij} \left\langle \boldsymbol{b}, \frac{\partial \boldsymbol{x}}{\partial \xi^j} \right\rangle . \tag{9.9}$$

Since

$$a_{ij} = \sum_k \frac{\partial x^k}{\partial \xi^i} \frac{\partial x^k}{\partial \xi^j} ,$$

it follows that

$$a^{ij} = \sum_k \frac{\partial \xi^i}{\partial x^k} \frac{\partial \xi^j}{\partial x^k} ,$$

and hence, from (9.9), that

$$\overline{b}^i = \sum_j \sum_k \frac{\partial \xi^i}{\partial x^k} \frac{\partial \xi^j}{\partial x^k} \left\langle \boldsymbol{b}, \frac{\partial \boldsymbol{x}}{\partial \xi^j} \right\rangle = \left\langle \boldsymbol{b}, \nabla \xi^i \right\rangle , \tag{9.10}$$

due to the inverse character of $\frac{\partial \xi^i}{\partial x^k}$ and $\frac{\partial x^k}{\partial \xi^i}$. Hence, in this case, the relation (9.8) reads

$$\boldsymbol{b} = \sum_i \left\langle \boldsymbol{b}, \nabla \xi^i \right\rangle \frac{\partial \boldsymbol{x}}{\partial \xi^i} , \tag{9.11}$$

and in the special case $\boldsymbol{b} = \nabla \xi^i$, we see that

$$\nabla \xi^i = \sum_j \sum_k \frac{\partial \xi^i}{\partial x^j} \frac{\partial \xi^k}{\partial x^j} \frac{\partial \boldsymbol{x}}{\partial \xi^k} , \quad i = 1, 2 . \tag{9.12}$$

In a completely analogous manner, we can compute the *covariant components* $\overline{b}_i$ of $\boldsymbol{b}$ in the expansion

$$\boldsymbol{b} = \overline{b}_1 \nabla \xi^1 + \overline{b}_2 \nabla \xi^2$$

as

$$\overline{b}_i = \left\langle \boldsymbol{b}, \frac{\partial x}{\partial \xi^i} \right\rangle .$$

Consequently, $\boldsymbol{b}$ can also be represented as

$$\boldsymbol{b} = \sum_i \left\langle \boldsymbol{b}, \frac{\partial \boldsymbol{x}}{\partial \xi^i} \right\rangle \nabla \xi^i , \tag{9.13}$$

and if we choose the special case $\boldsymbol{b} = \frac{\partial \boldsymbol{x}}{\partial \xi^i}$, we get

$$\frac{\partial \boldsymbol{x}}{\partial \xi^i} = \sum_j \sum_k \frac{\partial x^j}{\partial \xi^i} \frac{\partial x^j}{\partial \xi^k} \nabla \xi^k , \quad i = 1, 2 . \tag{9.14}$$

9.3 Metric Tensors

As we have seen, all geometric information about coordinate transformations between parameter space and physical space is already contained in the basis vectors $\frac{\partial x^i}{\partial \xi^j}$. The inner products of these vectors comprise the elements

$$g_{ij} = \left\langle \frac{\partial \boldsymbol{x}}{\partial \xi^i}, \frac{\partial \boldsymbol{x}}{\partial \xi^j} \right\rangle \tag{9.15}$$

of the *covariant metric tensor*

$$\boldsymbol{G} = (g_{i,j}) \,, \quad i,j = 1,2 \,.$$

This tensor is one of the fundamental entities in the differential geometry of coordinate transformations. Comparing (9.4) with (9.15), we immediately see that

$$\boldsymbol{G} = \boldsymbol{J} \cdot \boldsymbol{J}^T$$

and hence

$$g := \det \boldsymbol{G} = (\det \boldsymbol{J})^2 \,.$$

In analogy, there is also a *contravariant metric tensor* (g^{ij}) which is defined to be the inverse of the covariant metric tensor:

$$\sum_j g^{ij} g_{jk} = \delta_i^k = \begin{cases} 0; & i \neq k \\ 1; & i = k \end{cases} \,, \qquad i,k = 1,2 \,.$$

From this relation, we readily compute

$$g^{ij} = \langle \nabla \xi^i, \nabla \xi^j \rangle$$

and

$$\det(g^{ij}) = \frac{1}{g} \,.$$

Convenient formulae can now be derived which are quite useful. We give only a few of them and direct the reader to the literature.[148] The following relations are readily apparent:

$$g^{ij} = (-1)^{i+j} \cdot \frac{1}{g} \cdot g_{3-i\,3-j} \,, \quad i,j = 1,2$$

148) A book that is almost indispensable for practical work is J.F. Thompson, Z.U.A. Warsi, C.W. Mastin: *Numerical grid generation: foundations and applications.* Amsterdam: North-Holland 1985.

$$g_{ij} = (-1)^{i+j} \cdot g \cdot g^{3-i\,3-j}\,, \quad i,j = 1,2\,.$$

Looking back at (9.12) and (9.14), we see that our transformation relations between the two types of basis vectors can be written as

$$\frac{\partial \boldsymbol{x}}{\partial \xi^i} = \sum_k g_{ik} \nabla \xi^k\,, \quad i = 1,2$$

$$\nabla \xi^i = \sum_k g^{ik} \frac{\partial \boldsymbol{x}}{\partial \xi^k}\,, \quad i = 1,2\,.$$

9.4 Transforming Conservation Laws

We are now ready to adress the problem of transforming a given conservation law in physical space to one in parameter space. The derivatives of the Jacobian determinant $\det \boldsymbol{J} = \sqrt{g}$ form a key ingredient. Since

$$\det \boldsymbol{J} = \frac{\partial x^1}{\partial \xi^1}\frac{\partial x^2}{\partial \xi^2} - \frac{\partial x^1}{\partial \xi^2}\frac{\partial x^2}{\partial \xi^1}\,,$$

the computation of the derivative $\frac{\partial}{\partial \xi^k}$ is quite involved in terms of the number of terms which must be written down, but there are no conceptual difficulties. The result is

$$\frac{\partial}{\partial \xi^k} \det \boldsymbol{J} = (\det \boldsymbol{J}) \nabla_x \cdot \frac{\partial \boldsymbol{x}}{\partial \xi^k}\,,$$

where $\nabla_x\cdot$ denotes the divergence with respect to Cartesian coordinates. From this relation, we can derive the important formula[149)]

$$\sum_j \frac{\partial}{\partial \xi^j}\left((\det \boldsymbol{J})\frac{\partial \xi^j}{\partial \boldsymbol{x}}\right) = 0\,. \tag{9.16}$$

Consider a function $x^2 \ni (x^1, x^2) \mapsto f(x^1, x^2) \in \mathbb{R}$ defined in physical space. In order to compute a representation of the first derivatives of f in parameter space, we view f as $f(x^1(\xi^1, \xi^2), x^2(\xi^1, \xi^2))$, as above, and take derivatives. From what we derived above, it is clear that

$$\nabla_\xi f = \boldsymbol{J} \cdot \nabla_x f\,,$$

i.e.,

$$\nabla_x f = \boldsymbol{J}^{-1} \cdot \nabla_\xi f\,.$$

149) For a proof see, e.g., V.D. Liseikin: *Grid generation methods*. Berlin: Springer 1999, 46ff.

Now, since

$$J^{-1} = \frac{1}{\det J} \begin{bmatrix} \frac{\partial x^2}{\partial \xi^2} & -\frac{\partial x^1}{\partial \xi^2} \\ -\frac{\partial x^2}{\partial \xi^1} & \frac{\partial x^1}{\partial \xi^1} \end{bmatrix},$$

we obtain the following for the partial derivatives:

$$\frac{\partial f}{\partial x^1} = \frac{1}{\det J}\left(\frac{\partial x^2}{\partial \xi^2}\frac{\partial f}{\partial \xi^1} - \frac{\partial x^1}{\partial \xi^2}\frac{\partial f}{\partial \xi^2}\right) \tag{9.17}$$

$$\frac{\partial f}{\partial x^2} = \frac{1}{\det J}\left(-\frac{\partial x^2}{\partial \xi^1}\frac{\partial f}{\partial \xi^1} + \frac{\partial x^1}{\partial \xi^1}\frac{\partial f}{\partial \xi^2}\right). \tag{9.18}$$

The key ingredient of the transformation of conservation laws is the following theorem.

Theorem 9.1

$$\frac{\partial f}{\partial x^i} = \frac{1}{\det J}\sum_j \frac{\partial}{\partial \xi^j}\left((\det J)\frac{\partial \xi^j}{\partial x^i}f\right), \quad i = 1, 2 .$$

Proof: We compare the formulae (9.17) and (9.18) with the formula given in the lemma. Starting from the latter, we get

$$\begin{aligned} &\frac{1}{\det J}\sum_j \frac{\partial}{\partial \xi^j}\left((\det J)\frac{\partial \xi^j}{\partial x^i}f\right) \\ &= \frac{1}{\det J}\left[\frac{\partial}{\partial \xi^1}\left((\det J)\frac{\partial \xi^1}{\partial x^i}f\right) + \frac{\partial}{\partial \xi^2}\left((\det J)\frac{\partial \xi^2}{\partial x^i}f\right)\right] \\ &= \frac{1}{\det J}\left[\frac{\partial}{\partial \xi^1}\left((\det J)\frac{\partial \xi^1}{\partial x^i}\right)f + (\det J)\frac{\partial \xi^1}{\partial x^i}\frac{\partial f}{\partial \xi^1}\right. \\ &\quad \left. + \frac{\partial}{\partial \xi^2}\left((\det J)\frac{\partial \xi^2}{\partial x^i}\right)f + (\det J)\frac{\partial \xi^2}{\partial x^i}\frac{\partial f}{\partial \xi^2}\right]. \end{aligned}$$

Using (9.16), we see that the sum of the first and the third terms within the square brackets vanishes. Hence, we end up with

$$\frac{1}{\det J}\sum_j \frac{\partial}{\partial \xi^j}\left((\det J)\frac{\partial \xi^j}{\partial x^i}f\right) = \frac{\partial \xi^1}{\partial x^i}\frac{\partial f}{\partial \xi^1} + \frac{\partial \xi^2}{\partial x^i}\frac{\partial f}{\partial \xi^2} .$$

If we now replace the contravariant vectors with covariant ones according to the formulae we have already derived, we obtain (9.17) and (9.18) directly.

We can now give two examples.

Example

Consider the Laplace equation

$$\Delta f = 0 ,$$

which is obviously a conservation law, since $\Delta f = \nabla \cdot (\nabla f)$. Since

$$\Delta f = \sum_i \frac{\partial}{\partial x^i} \frac{\partial f}{\partial x^i} ,$$

we can replace the partial derivatives by means of Theorem 9.1 to yield

$$\begin{aligned} \frac{\partial}{\partial x^i} \frac{\partial f}{\partial x^i} &= \frac{1}{\det \boldsymbol{J}} \sum_j \frac{\partial}{\partial \xi^j} \left((\det \boldsymbol{J}) \frac{\partial \xi^j}{\partial x^i} \cdot \sum_k \frac{1}{\det \boldsymbol{J}} \frac{\partial}{\partial \xi^k} \left((\det \boldsymbol{J}) \frac{\partial \xi^k}{\partial x^i} f \right) \right) \\ &= \ldots \\ &= \frac{1}{\det \boldsymbol{J}} \sum_j \frac{\partial}{\partial \xi^j} \left((\det \boldsymbol{J}) \sum_k g^{kj} \frac{\partial f}{\partial \xi^k} \right) . \end{aligned}$$

Thus, the Laplace equation in the parameter space reads

$$\frac{1}{\det \boldsymbol{J}} \sum_j \frac{\partial}{\partial \xi^j} \left((\det \boldsymbol{J}) \sum_k g^{kj} \frac{\partial f}{\partial \xi^k} \right) = 0 .$$

Example

In our mathematical models of fluid flow, we frequently encounter derivatives of vector-valued flux functions in the form

$$\nabla \cdot \boldsymbol{f}(\boldsymbol{u}) = \sum_i \frac{\partial f^i}{\partial x^i} .$$

Using (9.16), we arrive at

$$\nabla \cdot \boldsymbol{f}(\boldsymbol{u}) = \frac{1}{\det \boldsymbol{J}} \sum_j \frac{\partial}{\partial \xi^j} \left((\det \boldsymbol{J} \cdot \bar{f}^j(\boldsymbol{u}) \right) , \tag{9.19}$$

in which $\bar{f}^j$ denotes the j-th contravariant component of the vector $\boldsymbol{f}$, i.e.,

$$\bar{f}^j(\boldsymbol{u}) = \sum_i f^i(\boldsymbol{u}) \frac{\partial \xi^j}{\partial x^i} .$$

Hence, the divergent part of a conservation law given in physical space transforms into (9.19) in parameter space. Note that this transformed term is again in conservation form.

We shall not describe second-order or higher derivatives, since the formulae follow along the lines developed here. Many books contain ready-to-use formulae[150] so that it is not necessary to work through tedious computations to arrive at transformed equations. Care must be taken to transform a conservation law in the physical domain into a *conservation* law in parameter space. It is much easier to transform a conservation law into an equation that does not possess the property of conservation.

9.5 Good Practice

We have expended quite some effort to show how differential equations can be transformed from physical space, in which the flow actually takes place, to parameter space, where finite difference methods can work. We have not, however, said a single word about the actual construction of the transformation, which is equivalent to the process of *grid generation*. In discretizing the transformed conservation laws, we now have to discretize the metric terms too.

In practice, one always thinks of the parameter domain as being discretized in the following way:

$$\begin{aligned} \xi^i &\in \left[1, 2, 3, \ldots, N^i\right] \\ \Delta\xi^i &= 1 \,. \end{aligned}$$

Hence, the discretized parameter space consists of integer values of the parameters with a spacing of 1. Depending on the computer language chosen, the parameters may take values $\xi^i \in 0, 1, 2, \ldots, N^i-1$, so that the numbering is somewhat arbitrary. This situation has already been shown in Fig. 9.1, where the grid lines and their corresponding straight lines in the parameter domain are depicted. For the metric terms, the most important of which are the derivatives $\frac{\partial \boldsymbol{x}}{\partial \xi^j}$, we can now choose any finite difference of choice. The most popular one used in production codes is the simple difference

$$\begin{aligned} \frac{\partial \boldsymbol{x}}{\partial \xi^1} &\approx \boldsymbol{x}\left(\xi^1+1, \xi^2\right) - \boldsymbol{x}\left(\xi^1, \xi^2\right) \\ \frac{\partial \boldsymbol{x}}{\partial \xi^2} &\approx \boldsymbol{x}\left(\xi^1, \xi^2+1\right) - \boldsymbol{x}\left(\xi^1, \xi^2\right) . \end{aligned}$$

Since the discrete parameter values are integers, it makes sense to write $\boldsymbol{x}(i,j)$ instead of $\boldsymbol{x}\left(\xi^1, \xi^2\right)$. The first step in the process of grid generation is to choose the *grid topology*. As an example, Fig. 9.2 shows a *C-type* grid around an airfoil-like obstacle, where the flow is assumed to come from the left.

150) See, e.g., J.F. Thompson, Z.U.A. Warsi, C.W. Mastin: *Numerical grid generation: foundations and applications.* Amsterdam: North-Holland 1985

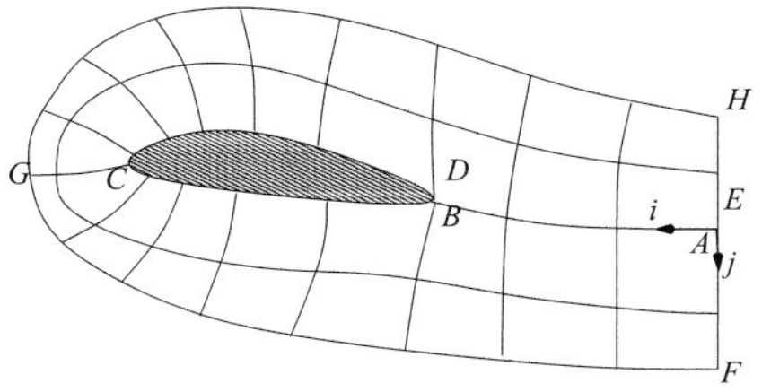

Fig. 9.2 C-type grid.

The corresponding parameter space is given in Fig. 9.3. It can be thought of as the physical space slit along the line AB and then continuously transformed into a rectangle. Note that the parts AB and DE of the parameter line $j = 1$ are logically identical. The surface of the obstacle corresponds to the line segment BCD, and the farfield is given by FGH. Note that $\boldsymbol{x}(1, 1)$ would correspond to A, $\boldsymbol{x}(4, 1)$ would correspond to B, and $\boldsymbol{x}(3, 1)$ to F.

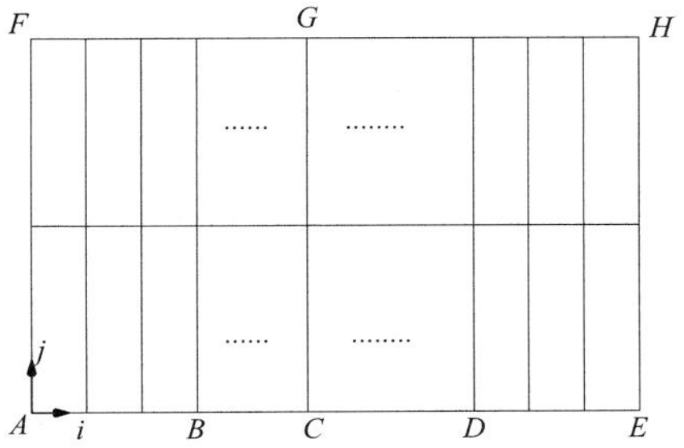

Fig. 9.3 C-type parameter space.

In contrast to the C-type topology, we could also have chosen the *O-type* grid, as shown in Fig. 9.4.

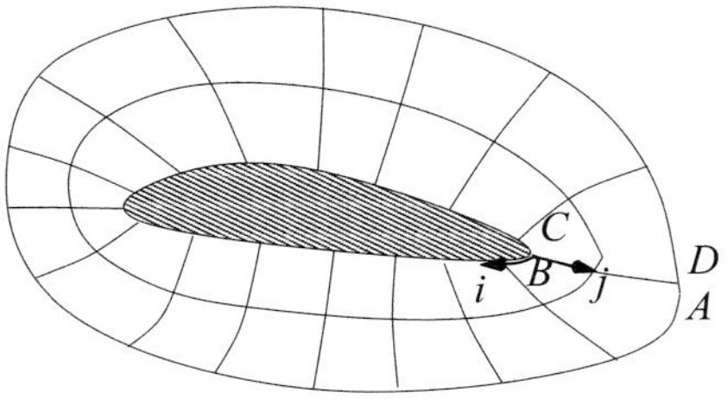

Fig. 9.4 O-type grid.

This type of grid topology is just a slight generalization of polar coordinates. Yet another topology is given by the *H-type* grid; see Fig. 9.5.

Here, the grid consists of two nearly Cartesian parts which fit logically together at the slit line $ABCDEFGH$. H-type grids are only used rarely for airfoils, because the grid quality near the leading and trailing edges of the airfoil is much worse than in the C- or O-type grids. By the *grid quality*, we mean the angles of the grid lines of the i-family with those of the j-family. A grid would be optimal if it was

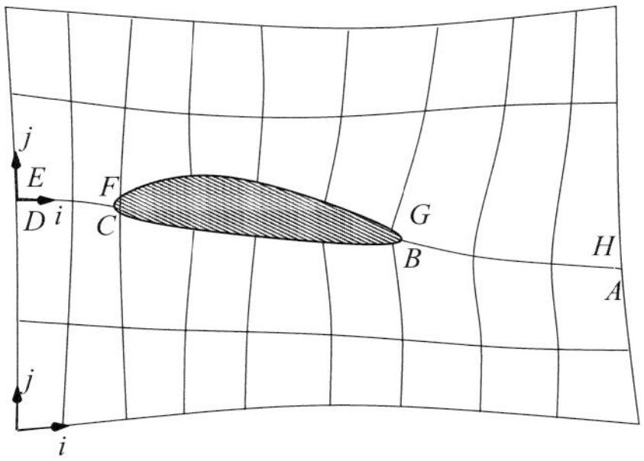

Fig. 9.5 H-type grid.

orthogonal everywhere, but this condition is seldom met. However, we should always try to get as close as possible to orthogonality, at least in regions sensible to the flow.

To describe only some of the hundreds of algorithms available for grid generation in detail would already be far beyond the goals of this book. The first step in grid generation is always to prescribe points on the surfaces of the obstacles in the flow and on the farfield. This is usually done with the help of spline functions or other interpolation techniques. Techniques used to generate the field grid, i.e., the grid points between the obstacle and farfield, can be divided into *algebraic* and *analytic* methods of grid generation. Interpolation techniques belong to the algebraic category. Stretching methods that allow a prescribed point density to be distributed on a line or plane can also be found in this category. Analytic methods compute grid points as solutions of partial differential equations or variational problems. The simplest partial differential equation used is the Laplace equation

$$\Delta \boldsymbol{\xi} = \mathbf{0}$$

for the parametric coordinates. Since we cannot solve for functions $\boldsymbol{\xi}(\boldsymbol{x})$ but we can for functions $\boldsymbol{x}(\boldsymbol{\xi})$ (these are our grid points), we need to transform the above equation along the lines outlined above to get

$$\sum_{i=1}^{2}\sum_{j=1}^{3} g^{ij}\frac{\partial^2 \boldsymbol{x}}{\partial \xi^i \partial \xi^j} = \mathbf{0} \; .$$

In terms of covariant metric coefficients instead of contravariant ones, we get

$$g_{22}\frac{\partial^2 \boldsymbol{x}}{\partial {\xi^1}^2} + g_{11}\frac{\partial^2 \boldsymbol{x}}{\partial {\xi^2}^2} - 2g_{12}\frac{\partial^2 \boldsymbol{x}}{\partial \xi^1 \partial \xi^2} = \mathbf{0} \; . \tag{9.20}$$

To solve for the position of the grid points, we have to prescribe an initial grid; this is usually derived via an algebraic method, i.e., interpolation. The second derivative that occurs is replaced by a finite difference, and the metric coefficients g_{ij} are also numerically computed by means of finite differences. Equation (9.20) is then a linear system of equations for the grid point positions $\boldsymbol{x}(i,j)$, which can be solved by an iterative method.

In order to analyze the properties of the grids generated by this system we consider the following case. We assume that $\xi^2 = 1$ describes the surface line of a two-dimensional convex body (consider it a circle), and $\xi^2 > 1$ are grid lines around that body but farther away. The $\xi^2 = \text{const}$ family is assumed to be a solution of the Laplace equation

$$\Delta\xi^2 = 0 ,$$

and that implicit function theorem allows the representation of $\xi^2(x^1, x^2) = \text{const}$ locally in the form $x_2 = f(x_1)$. Furthermore, we assume that

$$(x^2)'(0) = 0 \tag{9.21}$$

holds along the x^2-axis, i.e., that every grid line passes through the x^2-axis with a horizontal tangent. Let the grid line $\xi^2(x^1, x^2) = c_2$ pass through the point $(0, x^2_*)$, another line $\xi^2(x^1, x^2) = c_1$ pass through $(0, x^2_* - \Delta x^2)$, and a third one $\xi^2(x^1, x^2) = c_3$ pass through $(0, x^2_* + \Delta x^2)$, so that $c_1 < c_2 < c_3$. The implicitly defined curve $\xi^2(x^1, x^2) = \text{const}$ has a curvature given by

$$\kappa = \frac{\frac{\partial^2\xi^2}{\partial {x^1}^2}\left(\frac{\partial\xi^2}{\partial x^2}\right)^2 - 2\frac{\partial^2\xi^2}{\partial x^1\partial x^2}\frac{\partial\xi^2}{\partial x^1}\frac{\partial\xi^2}{\partial x^2} + \frac{\partial\xi^2}{\partial {x^2}^2}\left(\frac{\partial\xi^2}{\partial x^1}\right)^2}{\sqrt{\left(\left(\frac{\partial\xi^2}{\partial x^1}\right)^2 + \left(\frac{\partial\xi^2}{\partial x^2}\right)^2\right)^3}} . \tag{9.22}$$

Since the derivative in the explicit form $x^2 = f(x^1)$ is given by

$$(x^2)' = -\frac{\frac{\partial\xi^2}{\partial x^1}}{\frac{\partial\xi^2}{\partial x^2}} ,$$

we conclude from assumption (9.21) that $\frac{\partial\xi^2}{\partial x^1} = 0$ at every point on the x^2-axis. Substituting back into (9.22), we get the following for the curvature κ_0 of the grid lines at $x^1 = 0$:

$$\kappa_0 = \frac{\frac{\partial^2\xi^2}{\partial {x^1}^2}\left(\frac{\partial\xi^2}{\partial x^2}\right)^2}{\sqrt{\left(\frac{\partial\xi^2}{\partial x^2}\right)^6}} .$$

Due to the convexity of the surface line, it holds that $\kappa_0 > 0$ and therefore that

$$\frac{\partial^2\xi^2}{\partial {x^1}^2} > 0 .$$

However, since $\Delta\xi^2 = 0$ must be satisfied, we then obtain

$$\frac{\partial^2\xi^2}{\partial {x^2}^2} < 0 . \tag{9.23}$$

Replacing this last derivative by a finite difference of second order yields

$$\frac{\partial^2 \xi^2}{\partial x^{2\,2}} \approx \frac{\xi^2\left(0, x^2_* + \Delta x^2\right) - 2\xi^2\left(0, x^2_*\right) + \xi^2\left(0, x^2_* - \Delta x^2\right)}{(\Delta x^2)^2} ,$$

and it follows from (9.23) that

$$\xi^2\left(0, x^2_* + \Delta x^2\right) - \xi^2\left(0, x^2_*\right) < \xi^2\left(0, x^2_*\right) - \xi^2\left(0, x^2_* - \Delta x^2\right) ,$$

or, due to the numbering of the grid lines,

$$c_3 - c_2 < c_2 - c_1 .$$

Hence, in the case of a convex body, the number of grid lines is larger in the interval $\left[x^2_* - \Delta x^2, x^2_*\right]$ than in the interval $\left[x^2_*, x^2_* + \Delta x^2\right]$. Had we considered a concave surface, we would have ended up with

$$c_3 - c_2 > c_2 - c_1 .$$

Hence we can state the following:

Grid lines generated with system (9.20) move towards convex boundaries and tend to move away from concave ones.

A typical grid exhibiting the behavior demonstrated above can be seen in Fig. 9.6.

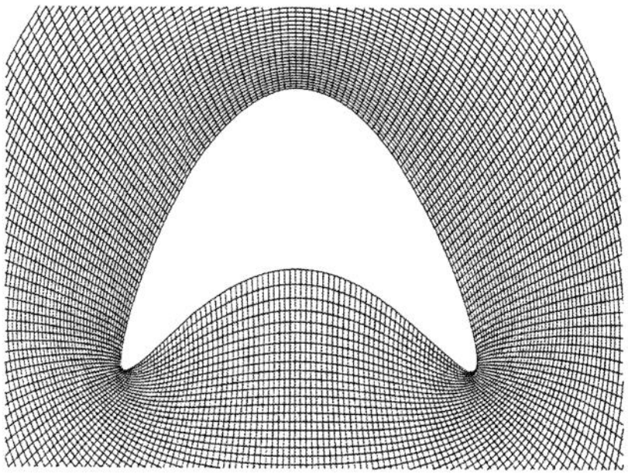

Fig. 9.6 Laplace grid around curved shape.

Equation (9.20) does not allow the user any kind of control over the grid line position. In order to remedy this, we can introduce two control functions or *source terms* Q^1 and Q^2 and solve the Poisson-type equation

$$\Delta\xi = \frac{1}{(\det J)^2}\begin{pmatrix} g^{11}Q^1 \\ g^{22}Q^2 \end{pmatrix} .$$

We could have used the simpler system $\Delta\xi = \begin{pmatrix} Q^1 \\ Q^2 \end{pmatrix} = \boldsymbol{Q}$ instead, but practice has shown that the contravariant metric coefficients used as factors as well as the square of the Jacobian determinant of the transformation add stability and robustness. In parameter space, this Poisson-type system reads

$$g_{22}\left(\frac{\partial^2 \boldsymbol{x}}{\partial \xi^{1\,2}} + Q^1\frac{\partial \boldsymbol{x}}{\partial \xi^1}\right) + g_{11}\left(\frac{\partial^2 \boldsymbol{x}}{\partial \xi^{2\,2}} + Q^2\frac{\partial \boldsymbol{x}}{\partial \xi^2}\right) - 2g_{12}\frac{\partial^2 \boldsymbol{x}}{\partial \xi^1 \partial \xi^2} = \boldsymbol{0} . \qquad (9.24)$$

If we now follow the analysis we presented for the Laplace equation for the Poisson equation $\Delta\xi = Q$ instead, we immediately see that negative values of Q^i force the grid lines of the ξ^i = const family towards decreasing values, and that positive values of Q^i force the grid lines of the ξ^i = const family towards increasing values. This principle is shown in Fig. 9.7.

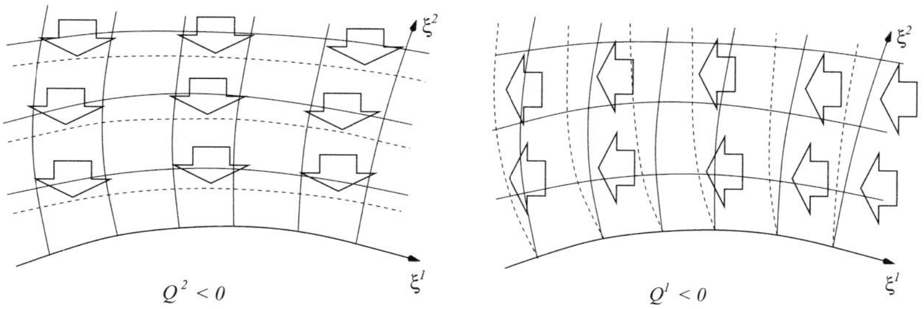

Fig. 9.7 Influence of the source terms Q^1 and Q^2.

There are several strategies for controlling the source terms Q^i, and we refer the reader to the literature.[151] As an example, we discuss one of the simplest but most successful strategies, which is called the *method of Hilgenstock*.[152] It is obvious from the discussion above that, at the boundaries of the physical domain, the angle of the outgoing grid lines as well as the spacing of the first grid line from the boundary can be influenced by the source terms. Let $\xi^2 = 1$ be the line of the boundary of interest (the airfoil, etc.). The direction ξ^2 is associated with the source term Q^2, and so Q^2 is responsible for the spacing between the grid lines of this family. Let d_r be the required spacing and

$$d = \left| \boldsymbol{x}\left(\xi^1, \xi^2 + 1\right) - \boldsymbol{x}\left(\xi^1, \xi^2\right) \right|$$

be the true spacing, which can be computed from the initial grid. If the difference $d_r - d$ vanishes, then no change in Q^2 is necessary; if $d_r - d > 0$, then the spacing of the given grid is too small and Q^2 has to grow. If $d_r - d < 0$, then the spacing of the grid is too large and Q^2 must decrease. The simplest function satisfying these requirements is

$$Q^2_{(k+1)} = Q^2_{(k)} + (d_r - d) ,$$

where (k) denotes an iteration level. In practice, this function can change its values drastically during the first few iterations, so that a better choice is

$$Q^2_{(k+1)} = Q^2_{(k)} + \arctan(d_r - d) \tag{9.25}$$

151) A good overview is provided in the book by Thomson, Warsi and Mastin cited before, but see also T. Sonar: *Grid generation using elliptic partial differential equations (Tech. Rep. 89-15)*. Braunschweig: DFVLR 1989.

152) A. Hilgenstock: *A method for the elliptic generation of three-dimensional grids with full boundary control (Int. Rep. 221-87 A 09)*. Braunschweig: DFVLR 1987 instead. Now consider a required angle α_r of the ξ^1 = const family at the boundary. The true angle can be computed from the given grid as

$$\alpha = \arccos\left(\frac{\left\langle \frac{\partial x}{\partial \xi^1}, \frac{\partial x}{\partial \xi^2} \right\rangle}{\left|\frac{\partial x}{\partial \xi^1}\right| \left|\frac{\partial x}{\partial \xi^2}\right|}\right).$$

If $\alpha_r - \alpha > 0$, then the actual angle is too small and Q^1 must decrease; if $\alpha_r - \alpha < 0$, then the angle is too large and Q^1 must increase. The simplest function implementing this is $Q^1_{(k+1)} = Q^1_{(k)} - (\alpha_r - \alpha)$, but the same remarks as above apply and

$$Q^1_{(k+1)} = Q^1_{(k)} - \arctan(\alpha_r - \alpha) \tag{9.26}$$

is a better choice. The typical grid generation cycle can then described as follows:

1. Let an initial grid be provided
2. Set $Q^1 = Q^2 = 0$ everywhere
3. Outer iteration: Set $k = 0$
   (a) Compute d and α at the boundary in the grid
   (b) Compute (9.25) and (9.26)
   (c) Inner iteration
      i. Solve (9.24) numerically via an iterative method
      ii. The result is a new grid given by its points $\boldsymbol{x}(i, j)$
   (d) Set $k = k + 1$
   (e) Continue outer iteration

A grid generated by this approach is shown in Fig. 9.8. The required angle at the boundary was chosen to be 90°.

Working with Laplace- or Poisson-type equations is known as *elliptic grid generation*, and has proven to be enormously successful in the past thirty years. These elliptic equations not only fit the boundary conditions in grid generation (prescribed point positions at the boundaries or even prescribed angles with the boundary) perfectly, but they permit very good control over the grid and also result in very smooth grids. There are, however, successful attempts to use parabolic or hyperbolic systems. In such cases, the point distribution must only be prescribed at the surface of the obstacle in the flow (this is the initial condition for the hyperbolic system), and the field points are then computed by marching away from the obstacle. It is not possible to prescribe the farfield.

In general, grid generation is still an art more than a science, and marvellous two- and three-dimensional grids can be found in the literature.

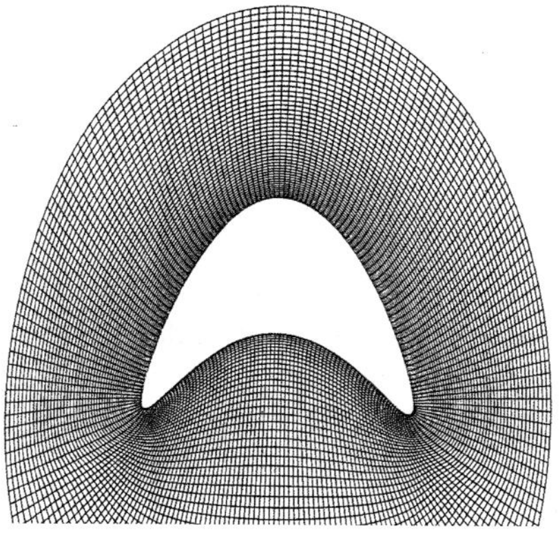

Fig. 9.8 Grid generated by an elliptic system with source term control.

9.6 Remarks Concerning Adaptation

The source terms Q^1 and Q^2 can be used not only to control grid spacing and angles at the boundary but also within the field. Hence, grid adaptivity can be achieved by carefully steering the source terms.

Suppose a function f on a one-dimensional grid gives an indication of some error,

$$w = f(x) .$$

The basic idea of adaptation is that the error should be equidistributed over the domain of w, i.e.,

$$w(x) \cdot \Delta x = \text{const} .$$

Upon introducing x as a function of ξ, this can be rewritten as

$$w(x(\xi))(x(\xi + \Delta\xi) - x(\xi)) = \text{const} ,$$

or it can be recast as

$$w(x(\xi))\frac{dx}{d\xi} = \text{const} .$$

If we differentiate this equation with respect to ξ, we get[153)]

$$w\frac{d^2w}{d\xi^2} + \frac{dw}{d\xi}\frac{dx}{d\xi} = 0 . \tag{9.27}$$

In the case of one dimension, our grid generation system (9.24) boils down to

$$\frac{d^2x}{d\xi^2} + Q\frac{dx}{d\xi} = 0 , \tag{9.28}$$

153) Note that there is now only one ξ, and so $\frac{d^2}{d\xi^2}$ means the second derivative with respect to ξ!

and if we compare this equation with (9.27), we see that

$$Q = -\frac{\frac{d^2x}{d\xi^2}}{\frac{dx}{d\xi}} . \qquad (9.29)$$

Now, from (9.27) we see that

$$\frac{\frac{d^2x}{d\xi^2}}{\frac{dx}{d\xi}} = -\frac{\frac{dw}{d\xi}}{w} ,$$

and inserting this into (9.29) results in

$$Q = \frac{\frac{dw}{d\xi}}{w} .$$

We have thus shown that, if we have an error function w, a grid can be generated by solving (9.28) with the above source term on which the error is equidistributed (which is more dense where the error is high).

We generalize this idea to two dimensions by simply setting the source terms to

$$Q^i = \frac{\frac{dw}{d\xi^i}}{w} , \quad i = 1, 2 . \qquad (9.30)$$

A very simple but efficient error function for applications in compressible flow is the function

$$w = 1 + |\nabla p| ,$$

in which the constant 1 serves to avoid a vanishing w if the pressure gradient is close to zero.

In practice, a good adaptation strategy is given by the following recipe. Keep the sources in the grid from the grid generation process, and call these sources Q^i_{old}. Compute the error function w and call the sources (9.30) Q^i_{adap}. Then, in the outer iteration of the grid generation process, compute

$$Q^i = Q^i_{\text{old}} + c \cdot Q^i_{\text{adap}}$$

and generate a grid with these new source terms. The constant c is a damping factor ≤1 that is used to keep the inner iteration stable.

10
Finite Volume Models

10.1
Difference Methods on Unstructured Grids

We have already discussed finite volume methods in Section 7.8, and we arrived at Eq. (7.39):

$$\frac{1}{\Delta t}\int_{\sigma_i}\{u^{n+1}(x,y)-u^n(x,y)\}\,\mathrm{d}\sigma_i+\sum_{k\in I_i}\int_{\partial\sigma_{i,k}}\left\langle f(u^n),\boldsymbol{n}^{i,k}\right\rangle\,\mathrm{d}s=\mathcal{O}(\Delta t) \tag{10.1}$$

where $\boldsymbol{n}^{i,k}$ denotes the outer unit normal vector at the edge $\partial\sigma_{i,k}$, and where we have used a simple forward difference in time.[154] Writing

$$v_i^0=\frac{1}{|\sigma_i|}\int_{\sigma_i}u_0(x,y)\,\mathrm{d}\sigma_i$$

for the cell average of the initial functions, neglecting the error term in (10.1), and writing $s_{i,k}$ for the length of the edge $\partial s_{i,k}$ resulted in (7.40):

$$v_i^{n+1}=v_i^n-\frac{\Delta t}{|\sigma_i|}\sum_{k\in I_i}\left\langle f(\tilde{v}_{i,k}^v),\boldsymbol{n}^{i,k}\right\rangle s_{i,k}\ .$$

Evaluating the inner product on certain polygonal cells σ_i will result in a finite difference scheme due to the signs of the components of the normal vector. Since the kind of volumes σ_i used is arbitrary – the σ_i may be a quadrilateral, a triangle or any other polygonal shape – it is natural from a finite difference point of view to call finite volume schemes *finite difference methods on unstructured grids*, and they have been called this in some of the literature.[155] In 2007, a survey work on finite volume methods was published[156] where the authors exploited a slightly different point of view: finite volume methods can also be interpreted as a kind of discontinuous Galerkin FEM.

154) This by no means necessary; any time-stepping scheme can be used.

155) See, for example, B. Heinrich: *Finite difference methods on irregular networks.* Stuttgart: Birkhäuser 1987

156) K.W. Morton, T. Sonar: Acta Num. **16** (2007) 155–238

Mathematical Models of Fluid Dynamics. R. Ansorge and T. Sonar

ISBN: 978-3-527-40774-3

For the sake of simplicity, we will employ the finite volume approach for regular triangulations in the plane. In general, there are three degrees of freedom for the type of discretization when a finite volume method is being set up, namely a cell-center approach, a cell-vertex approach, and a node-centered approach.

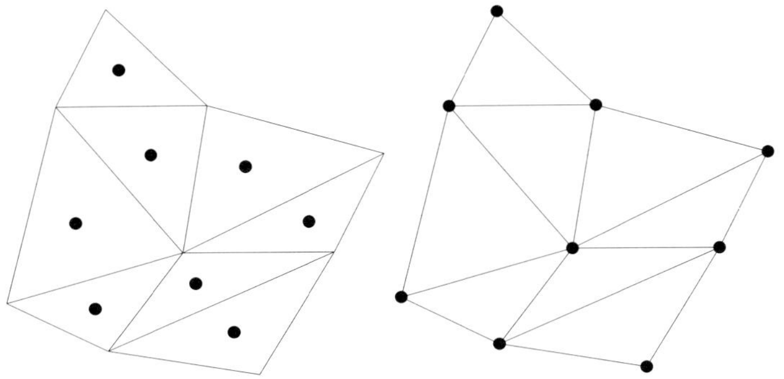

Fig. 10.1 Cell-centered and cell-vertex approach.

The approaches differ in terms of the locations of the unknowns, and hence lead to different cells or *control volumes* σ_i. We have found the node-centered formulation to give the most pleasing results. In this approach, we consider the *barycentric subdivision* of triangles in wich the barycenter of the triangle itself is connected to the barycenters (i.e., midpoints) of the edges via straight lines.

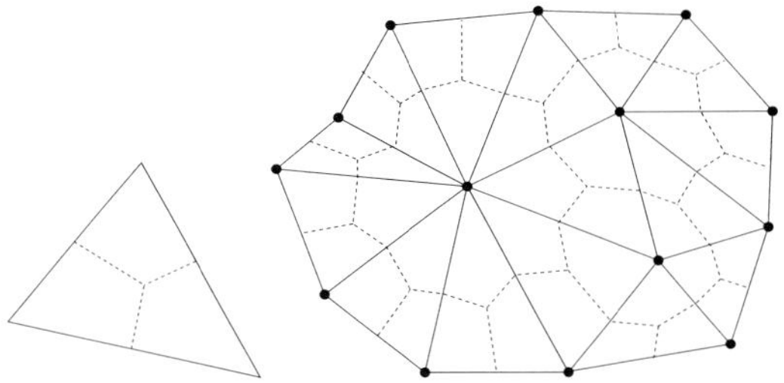

Fig. 10.2 Barycentric subdivision and the resulting primary and secondary grids.

The barycentric subdivisions build a new cell around each node: the control volume σ_i belonging to node i of the triangulation. While the triangulation is called the *primary grid*, we call the set of all control volumes the *secondary grid*. The geometry between two control volumes, which are also called *boxes*, is shown in Fig. 10.3. Due to the barycentric subdivision of the triangles, the box boundary now consists of two parts ∂s_{ik}^1 and ∂s_{ik}^2 with lengths s_{ik}^1 and s_{ik}^2, respectively, which give rise to two outer (with respect to σ_i) unit normal vectors $\boldsymbol{n}_1^{ik}$ and $\boldsymbol{n}_2^{ik}$. Hence, we can rewrite (10.1) in the form

$$\frac{\mathrm{d}}{\mathrm{d}t}\frac{1}{|\sigma_i|}\int_{\sigma_i} u \,\mathrm{d}\sigma_i = -\frac{1}{|\sigma_i|}\sum_{k\in I_i}\sum_{\ell=1}^{2}\int_{\partial\sigma_{ik}^{\ell}}\left\langle \boldsymbol{f}(u), \boldsymbol{n}_\ell^{ik}\right\rangle \mathrm{d}s\,, \tag{10.2}$$

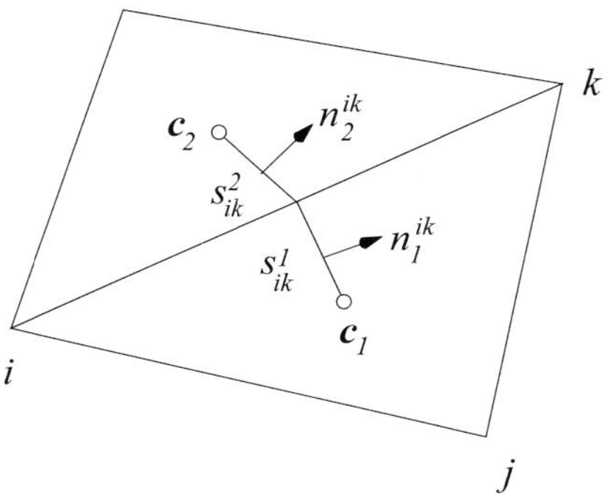

Fig. 10.3 Geometry between two boxes.

where we have allowed for a more general time-stepping scheme. It is immediately apparent from this formulation that the finite volume methods are numerical schemes for computing cell averages over the boxes. In order to give us some freedom in terms of the accuracy of the resulting scheme, we now introduce a parametrization of each of the boundary segments $\partial\sigma_{ik}^{\ell}, \ell = 1, 2$ in the form

$$[-1, 1] \ni \zeta \mapsto \boldsymbol{x}_{ik}^{\ell}(\zeta) = \frac{1}{2}\left(\boldsymbol{c}_{\ell} + \frac{1}{2}(\boldsymbol{x}_i + \boldsymbol{x}_k)\right) + \frac{\zeta}{2}\left(\frac{1}{2}(\boldsymbol{x}_i + \boldsymbol{x}_k) - \boldsymbol{c}_{\ell}\right) ,$$

where $\boldsymbol{c}_1$ and $\boldsymbol{c}_2$ are the coordinates of the barycenters of the two adjacent triangles sharing edge ik, and $\boldsymbol{x}_i$, $\boldsymbol{x}_k$ denote the coordinates of the points i and k of the triangulation, respectively. Aided by this parametrization, we are now able to introduce Gaussian quadrature points on $\partial\sigma_{ik}$, so that (10.2) now reads as

$$\frac{\mathrm{d}}{\mathrm{d}t}\frac{1}{|\sigma_i|}\int_{\sigma_i} u \,\mathrm{d}\sigma_i = -\frac{1}{|\sigma_i|}\sum_{k\in I_i}\sum_{\ell=1}^{2}\frac{s_{ik}^{\ell}}{2}\left[\sum_{\nu=1}^{n_G}\omega_{\nu}^{\ell}\left\langle \boldsymbol{f}\left(u\left(\boldsymbol{x}_{ik}^{\ell}(\zeta_{\nu}), t\right)\right), \boldsymbol{n}_{\ell}^{ik}\right\rangle + \mathcal{O}(h^{2n_G})\right] \tag{10.3}$$

where n_G denotes the number of Gaussian points, ω_{ν} corresponds to the weights of the quadrature rule, and the error includes a typical length measure h of the triangulation involved.

We now introduce a numerical flux function, as we did in the derivation of (7.44). In that case we included s_{ik} in the definition of the numerical flux $g_{i,k}$ because this was convenient in the derivation of a low-order finite volume scheme employing only one Gaussian point. Here we redefine our definition, and we want to introduce a numerical flux in the form

$$(u_l, u_r; \boldsymbol{n}) \mapsto g(u_l, u_r; \boldsymbol{n}) ,$$

which is required to satisfy the consistency condition

$$g(w, w; \boldsymbol{n}) = \langle \boldsymbol{f}(w), \boldsymbol{n}\rangle .$$

When compared to (7.43), it is obvious that only s_{ik} is excluded from our redefinition. Introducing a numerical flux function of this kind into (10.3) yields

$$\frac{\mathrm{d}}{\mathrm{d}t}\frac{1}{|\sigma_i|}\int_{\sigma_i} u \,\mathrm{d}\sigma_i$$

$$= -\frac{1}{|\sigma_i|} \sum_{k \in I_i} \sum_{\ell=1}^{2} \frac{s_{ik}^{\ell}}{2} \left[\sum_{\nu=1}^{n_G} \omega_\nu^\ell g\left(u\left(\boldsymbol{x}_{ik}^\ell(\zeta_\nu), t \right), u\left(\boldsymbol{x}_{ik}^\ell(\zeta_\nu), t \right); \boldsymbol{n}_\ell^{ik} \right) + \mathcal{O}(h^{2n_G}) \right] . \tag{10.4}$$

10.2
Order of Accuracy and Basic Discretization

We have not yet mentioned how we intend to evaluate u at different Gaussian points. At the moment, we are only able to define a first-order finite volume method based entirely on cell averages, which we call *basic discretization*. If v_i denotes the cell average of our numerical solution (which still depends on time!), where we employ the discretization of the initial values as

$$v_i(t=0) = \frac{1}{|\sigma_i|} \int_{\sigma_i} u_0(\boldsymbol{x}, t) \, \mathrm{d}\sigma_i ,$$

we can evaluate the numerical flux function at exactly one Gaussian point – in the middle of ∂s_{ik}^ℓ between the cell averages v_i and v_k. The parameter value of the middle point is $\zeta_\nu = 0$, and the corresponding weight is given by $\omega_\nu^\ell = 2$. Hence, the basic discretization is given by

$$\frac{\mathrm{d}}{\mathrm{d}t} v_i(t) = -\frac{1}{|\sigma_i|} \sum_{k \in I_i} \sum_{\ell=1}^{2} s_{ik}^\ell g\left(v_i(t), v_k(t); \boldsymbol{n}_\ell^{ik} \right) . \tag{10.5}$$

Note that this is exactly the same scheme given in (7.44) (rewritten for our boxes) if forward difference is applied for the time derivative.

Concerning the spatial accuracy, we define *order* as follows. Our basic discretization (10.5) is said to be *of order r* if

$$g\left(v_i, v_k; \boldsymbol{n}_\ell^{ik} \right) = \left\langle \boldsymbol{f}(v(\boldsymbol{x}_{ik}(0), t)), \boldsymbol{n}_\ell^{ik} \right\rangle + \mathcal{O}(h^r) .$$

Since we have only used one Gaussian point ($n_G = 1$), the only possible values for r are 1 or 2. In order to decide on the accuracy question, we must examine the approximation properties of cell averages. From the definition of a barycenter, we know that

$$\frac{1}{|\sigma_i|} \int_{\sigma_i} \boldsymbol{x} \, \mathrm{d}\sigma_i = \boldsymbol{c}_i \tag{10.6}$$

if $\boldsymbol{c}_i$ are the coordinates of the barycenters of σ_i. Now let $\boldsymbol{\alpha} = (\alpha_1, \alpha_2)$ be a multi-index and consider the Taylor expansion of a smooth function u around the barycenter $\boldsymbol{c}_i$:

$$u(\boldsymbol{x}, t) = \sum_{\mu=0}^{r-1} \frac{1}{\mu!} \sum_{|\alpha|=\mu} (\boldsymbol{x} - \boldsymbol{c}_i)^{\boldsymbol{\alpha}} \, \partial^{\boldsymbol{\alpha}} u\big|_{\boldsymbol{x}=\boldsymbol{c}_i} + \mathcal{O}(|\boldsymbol{x} - \boldsymbol{c}_i|^r) .$$

Applying the cell average results in

$$\frac{1}{|\sigma_i|}\int_{\sigma_i} u \, d\sigma_i = \sum_{\mu=0}^{r-1} \frac{1}{\mu!} \sum_{|\alpha|=\mu} \frac{1}{|\sigma_i|}\int_{\sigma_i} (x - c_i)^\alpha \, d\sigma_i \; \partial^\alpha u\big|_{x=c_i} + \mathcal{O}(|x - c_i|^r) \, ,$$

and choosing $r = 2$ yields

$$\frac{1}{|\sigma_i|}\int_{\sigma_i} u \, d\sigma_i = u(c_i, t) + \left\langle \frac{1}{|\sigma_i|}\int_{\sigma_i} (x - c_i) \, d\sigma_i, \nabla_x u(x, t)\big|_{x=c_i} \right\rangle + \mathcal{O}(|x - c_i|^2) \, .$$

Due to (10.6), we find that $\int_{\sigma_i} (x - c_i) \, d\sigma_i = 0$, and we finally arrive at

$$\frac{1}{|\sigma_i|}\int_{\sigma_i} u \, d\sigma_i = u(c_i, t) + \mathcal{O}(h^2) \, .$$

Hence, the cell average operator represents the function u at the barycenter with second-order accuracy. However, from the computation above, it follows that, if we take another point $x_i \in \sigma_i$ which is *not* the barycenter, then

$$\frac{1}{|\sigma_i|}\int_{\sigma_i} u \, d\sigma_i = u(x_i, t) + \mathcal{O}(h)$$

holds. We can thus conclude that our basic discretization (10.5) is of first-order spatial accuracy only.

10.3 Higher-Order Finite Volume Schemes

The analysis of accuracy given above can provide us with a hint about how to increase the spatial accuracy of finite volume methods. Suppose there were a smooth function

$$(x, t) \mapsto p_i(x, t)$$

defined for all $x \in \sigma_i$ for which

$$p_i(x, t) - u(x, t) = \mathcal{O}(h^r)$$

for all $x \in \sigma_i$, as well as

$$\frac{1}{|\sigma_i|}\int_{\sigma_i} p_i \, d\sigma_i = \frac{1}{|\sigma_i|}\int_{\sigma_i} u \, d\sigma_i \, .$$

Then we could define a finite volume method as

$$\frac{d}{dt}\frac{1}{|\sigma_i|} v_i = -\frac{1}{|\sigma_i|}\sum_{k\in I_i}\sum_{\ell=1}^{2}\frac{s_{ik}^\ell}{2}\sum_{\nu=1}^{n_G}\omega_\nu^\ell g\left(p_i\left(x_{ik}^\ell(\zeta_\nu), t\right), p_k\left(x_{ik}^\ell(\zeta_\nu), t\right); n_\ell^{ik}\right) . \tag{10.7}$$

We say that finite volume method (10.7) is *of spatial order q* if

$$\sum_{\nu=1}^{n_G} \omega_\nu^\ell g\left(p_i\left(\boldsymbol{x}_{ik}^\ell(\zeta_\nu), t\right), p_k\left(\boldsymbol{x}_{ik}^\ell(\zeta_\nu), t\right); \boldsymbol{n}_\ell^{ik}\right)$$
$$= \int_{-1}^{1} \left\langle \boldsymbol{f}\left(u\left(\boldsymbol{x}_{ik}^\ell(\zeta), t\right)\right), \boldsymbol{n}_\ell^{ik} \right\rangle \mathrm{d}\zeta + \mathcal{O}(h^q) .$$

It is immediately apparent that this definition generalizes that given for the basic discretization where we used $n_G = 1$ and, correspondingly, $\omega_1^\ell = 2$. It is then easy to show (see [20]) that a finite volume method (10.7) is of order $q = \min\{r, 2n_G\}$, where r is determined by

$$g\left(p_i\left(\boldsymbol{x}_{ik}^\ell(\zeta_\mu), t\right), p_k\left(\boldsymbol{x}_{ik}^\ell(\zeta_\mu), t\right); \boldsymbol{n}_\ell^{ik}\right) = \left\langle \boldsymbol{f}\left(u\left(\boldsymbol{x}_{ik}^\ell(\zeta_\mu), t\right)\right), \boldsymbol{n}_\ell^{ik} \right\rangle + \mathcal{O}(h^r) . \quad (10.8)$$

We are now left with the question of how to derive functions p_i on the boxes σ_i. This task is usually known as the *recovery procedure*.

Before we look at recovery procedures, we will make a few remarks about time-stepping schemes. We write any finite volume scheme in the abstract form

$$\frac{\mathrm{d}}{\mathrm{d}t} v_i(t) = -\mathcal{L}^h(\{\boldsymbol{p}(t)\}) ,$$

where $v_i(t)$ is the cell average on box σ_i at time t, and $\boldsymbol{p}(t)$ represents the recovery polynomial to be computed. The operator $\mathcal{L}^h$ stands for the complete right hand side of a finite volume method. We will only discuss explicit methods in this context. However, it is worth noting that Meister has done groundbreaking work on implicit time-stepping schemes, which can be found in the literature.[157]

The simplest explicit time-stepping method is the Euler forward difference:

$$\frac{\mathrm{d}}{\mathrm{d}t} v_i(t) = \frac{v_i(t + \Delta t) - v_i(t)}{\Delta t} + \mathcal{O}(\Delta t) .$$

This method is first-order accurate in time and exhibits a small range of stability, so that in actual computations of fluid flow, the restriction on the time step Δt is often more restrictive than that predicted by the CFL condition. This said, it must also be admitted that the forward difference method is the first choice when developing and testing a new method.

Very popular in engineering circles is the classical four-step Runge–Kutta time-stepping method, but it is not advisable to use it with the kinds of schemes we are after. This is because temporal errors from the Runge–Kutta scheme can turn a finite volume method with a spatial total variation diminishing operator $\mathcal{L}^h$ into a non-monotone scheme. Shu and Osher developed families of Runge–Kutta schemes which conform with the monotonicity of the spatial discretization.[158] We

157) See A. Meister: J. Comp. Phys. **140** (1998) 311–345

158) See C.-W. Shu, S. Osher: J. Comp. Phys. **77** (1988) 439–417, and (the second part of that work): J. Comp. Phys. **83** (1989) 32–78

can highly recommend the use of the second- or third-order variant, for which there is an economical form given by

$$\begin{aligned}
v_i^{(0)} &= v_i(n\Delta t)\\
v_i^{(1)} &= v_i^{(0)} - \Delta t \mathcal{L}^h\left(\boldsymbol{p}^{(0)}\right)\\
v_i^{(2)} &= v_i^{(0)} - \frac{1}{2}\Delta t \mathcal{L}^h\left(\boldsymbol{p}^{(0)}\right) - \frac{1}{2}\Delta t \mathcal{L}^h\left(\boldsymbol{p}^{(1)}\right)\\
v_i((n+1)\Delta t) &= v_i^{(2)}
\end{aligned}$$

for the second-order variant and

$$\begin{aligned}
v_i^{(0)} &= v_i(n\Delta t)\\
v_i^{(1)} &= v_i^{(0)} - \Delta t \mathcal{L}^h\left(\boldsymbol{p}^{(0)}\right)\\
v_i^{(2)} &= v_i^{(0)} - \frac{1}{4}\Delta t \mathcal{L}^h\left(\boldsymbol{p}^{(0)}\right) - \frac{1}{4}\Delta t \mathcal{L}^h\left(\boldsymbol{p}^{(1)}\right)\\
v_i^{(3)} &= v_i^{(0)} - \frac{1}{6}\Delta t \mathcal{L}^h\left(\boldsymbol{p}^{(0)}\right) - \frac{1}{6}\Delta t \mathcal{L}^h\left(\boldsymbol{p}^{(1)}\right) - \frac{2}{3}\Delta t \mathcal{L}^h\left(\boldsymbol{p}^{(2)}\right)\\
v_i((n+1)\Delta t) &= v_i^{(3)}
\end{aligned}$$

for the third-order scheme. Here we have used the notation $v_i^{(k)}$ to mark the stages of the time-stepping scheme. It is worth noting that, during the time stepping, one additional recovery process is required for the second-order scheme while two other recoveries must be performed in the third-order case. Since the recovery has to be performed for each box and is usually time-consuming, the second-order scheme may be the practical choice.

10.4 Polynomial Recovery

At the beginning of each new time step, we would like to construct on each box σ_i a polynomial p_i of fixed degree such that both

$$\frac{1}{|\sigma_i|}\int_{\sigma_i} p_i \,\mathrm{d}\sigma_i = \frac{1}{|\sigma_i|}\int_{\sigma_i} u \,\mathrm{d}\sigma_i \,,$$

and

$$p_i(\boldsymbol{x}, t) - u(\boldsymbol{x}, t) = \mathcal{O}(h^r)$$

hold for each fixed time t. Polynomials satifying these conditions are called *recovery polynomials*. They can always be written in the form

$$p_i(\boldsymbol{x}, t) = \sum_{\mu=1}^{r-1}\frac{1}{\mu!}\sum_{|\alpha|=\mu}(\boldsymbol{x} - c_i)^{\alpha} a_{\alpha}(t) \,, \tag{10.9}$$

where $c_i = (c_{1,i}, c_{2,i})$ is again the barycenter of box σ_i; i.e., a linear recovery polynomial ($r = 2$) would be

$$p_i(x, t) = a_{(0,0)} + a_{(1,0)}(x - c_{1,i}) + a_{(0,1)}(y - c_{2,i})$$

and a quadratic recovery polynomial would read

$$\begin{aligned} p_i(x, t) &= a_{(0,0)} + a_{(1,0)}(x - c_{1,i}) + a_{(0,1)}(y - c_{2,i}) \\ &\quad + \frac{1}{2} a_{(2,0)}(x - c_{1,i})^2 + a_{(1,1)}(x - c_{1,i})(y - c_{2,i}) + \frac{1}{2} a_{(0,2)}(y - c_{2,i})^2 . \end{aligned}$$

It is now obvious that the coefficients a_α of a recovery polynomial (10.9) must satisfy the condition

$$a_\alpha = \partial^\alpha u\big|_{x=c_i} + \mathcal{O}\left(h^{r-|\alpha|}\right),$$

since if u is smooth and can be expanded according to

$$u(x, t) = \sum_{\mu=1}^{r-1} \frac{1}{\mu!} \sum_{|\alpha|=\mu} (x - c_i)^\alpha \, \partial^\alpha u\big|_{x=c_i} + \mathcal{O}(h^r) ,$$

then

$$p_i(x, t) - u(x, t) = \sum_{\mu=1}^{r-1} \frac{1}{\mu!} \sum_{|\alpha|=\mu} (x - c_i)^\alpha \left(a_\alpha - \partial^\alpha u\big|_{x=c_i}\right) + \mathcal{O}(h^r) .$$

The condition on the coefficients then follows from

$$\begin{aligned} |p_i(x, t) - u(x, t)| &\le \sum_{\mu=1}^{r-1} \frac{1}{\mu!} \sum_{|\alpha|=\mu} |x - c_i|^{|\alpha|} \left|a_\alpha - \partial^\alpha u\big|_{x=c_i}\right| + \mathcal{O}(h^r) \\ &\le \sum_{\mu=1}^{r-1} \frac{1}{\mu!} \sum_{|\alpha|=\mu} h^{|\alpha|} \left|a_\alpha - \partial^\alpha u\big|_{x=c_i}\right| + \mathcal{O}(h^r) . \end{aligned}$$

It is clear that a recovery polynomial of degree 1 ($r = 2$) has two unknown coefficients, while a polynomial of degree 2 ($r = 3$) has six unknown coefficients. In general, in two spatial dimensions, the number of coefficients to be computed for order r is given by the binomial coefficient

$$\binom{r+1}{r-1} .$$

To compute the unknown coefficients of a recovery polynomial (10.9) on box σ_i, one needs the cell average v_i on σ_i, as well as $\binom{r+1}{r-1} - 1$ further cell averages on boxes in the neighborhood of σ_i. A recovery algorithm must therefore search for neighboring boxes and then compute the polynomial.

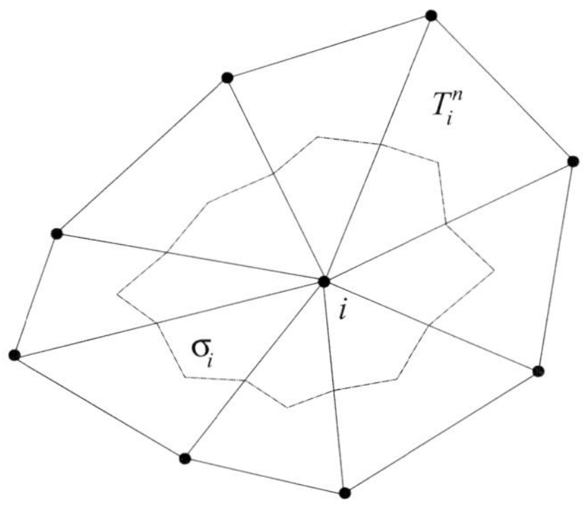

Fig. 10.4 Geometry for linear recovery.

We start with a simple but robust and accurate algorithm. Consider the box σ_i around node i, like the one in Fig. 10.4.

Since the data is stored at the nodes of the triangulation, we can compute a linear polynomial on every triangle T_i^n sharing node i:

$$\pi_{T_i^n} = \sum_{m=1}^{3} v_m \Phi_m^n$$

where

$$\Phi_m^n(\boldsymbol{x}, t) = b_0^m(t) + b_1^m(t)x + b_2^m(t)y$$

is the Lagrangian interpolation polynomial defined by

$$\Phi_m^n(\boldsymbol{x}_l, t) = \delta_i^l ,$$

where the $\boldsymbol{x}_l$ denote the nodes of triangle T_i^n. Hence, we end up with as many linear polynomials $\pi_{T_i^n}, n = 1, \ldots, \#T$ as there are triangles sharing node i (and which therefore constitute the box σ_i). Note that these polynomials are interpolants of cell averages.

A linear recovery polynomial should take the form

$$p_i(\boldsymbol{x}, t) = a_{(0,0)} + a_{(1,0)}(x - c_{1,i}) + a_{(0,1)}(y - c_{2,i}) ,$$

and since $p_i(\boldsymbol{c}_i, t) = v_i(t)$, the value of $a_{(0,0)}$ is the cell average on σ_i. Hence, we can rewrite the polynomial as

$$p_i(\boldsymbol{x}, t) = v_i + \langle \nabla w_i, \boldsymbol{x} - \boldsymbol{c}_i \rangle , \tag{10.10}$$

where we have written $\nabla w_i = (a_{(1,0)}, a_{(0,1)})$ on σ_i. We can now use our linear interpolants on the triangles surrounding i and define

$$\nabla w_i := \frac{1}{|\sigma_i|} \int_{\sigma_i} \nabla_{\boldsymbol{x}} \pi_{T_i^n} \Big|_{T_i^n} \, \mathrm{d}\sigma_i .$$

Since the gradients $\nabla_x \pi_{T_i^n}$ are constant on the triangles they belong to, we could also have written

$$\nabla w_i = \frac{1}{|\sigma_i|} \sum_n \nabla_x \pi_{T_i^n} |T_i^n \cap \sigma_i| \ .$$

This would indeed result in a linear recovery polynomial, but it would result in serious stability problems if applied to hyperbolic conservation laws. A cure is provided by direct *slope limiting*. The recovery polynomial (10.10) is augmented by a slope limiter L_i:

$$p_i(x, t) = v_i + L_i \langle \nabla w_i, x - c_i \rangle \ .$$

Several functions have been derived in the literature, and one that is widely used is that designed by Barth and Jespersen:[159)]

$$L_i = \min_{\partial \sigma_{ik}^\ell} L_{\partial \sigma_{ik}^\ell}$$

with

$$L_{\partial \sigma_{ik}^\ell} = \begin{cases} \min \left\{ 1, \dfrac{v_{ik}^+ - v_i}{\hat{p}_i - v_i} \right\} ; & \hat{p}_i - v_i > 0 \\ \min \left\{ 1, \dfrac{v_{ik}^- - v_i}{\hat{p}_i - v_i} \right\} ; & \hat{p}_i - v_i < 0 \\ 1 \qquad ; & \hat{p}_i - v_i = 0 \end{cases}$$

where $v_{ik}^+ = \max_{k \in I_i} \{v_i, v_k\}$, $v_{ik}^- = \min_{k \in I_i} \{v_i, v_k\}$, and $\hat{p}_i$ denotes the evaluation of the unlimited polynomial (10.10) at the Gaussian points on $\partial \sigma_{ik}^\ell$. In practice, however, slight oscillations can still be seen in the computation of shock waves. The scheme can be made total variation diminishing if the resulting recovery polynomial on every box is also checked to make sure that it respects the monotonicity of the cell averages.

An even simpler algorithm is given by the following recipe. As we did above, we compute the different Lagrange polynomials $\pi_{T_i^n}$ and take their gradients. Then we choose

$$\nabla w_i = \min_k |\nabla_x \pi_{T_i^k}|$$

for the gradient of the recovery polynomial on box σ_i.

We now want to describe an algorithm which is capable of recovering higher-degree polynomials. It was developed by Abgrall[160)] and is exploited in [20]. As we

159) T.J. Barth, D.C. Jespersen: *The design and application of upwind schemes on unstructured meshes (AIAA Paper 89-03660).* Reston: AIAA 1989

160) R. Abgrall: J. Comp. Phys. **114** (1994) 45–58

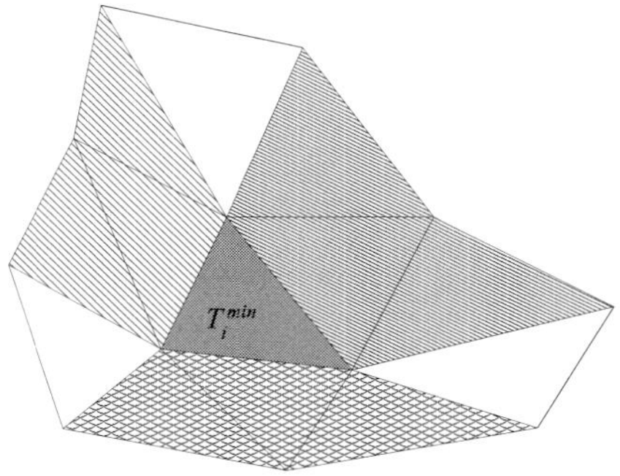

Fig. 10.5 Three sets of triangles for quadratic recovery.

did above, we construct linear polynomials $\pi_{T_i^n}$ on every triangle T_i^n sharing the node i via the ansatz

$$\pi_{T_i^n} = \sum_{|\alpha|\le 1} (x - c_i)^{\alpha} a_{\alpha}^n$$

and the coefficients from

$$\frac{1}{|\sigma_l|} \int_{\sigma_l} \pi_{T_i^n} \, \mathrm{d}\sigma_l = v_l \ ,$$

where the index l runs over the three nodes of the triangle. Among all those triangles sharing node i, we denote the triangle for which

$$\sum_{|\alpha|=1} |a_{\alpha}^N| = \min_n \left\{ \sum_{|\alpha|=1} |a_{\alpha}^n| \right\}$$

is valid as $T_i^{\min}$ (T_i^N, say). If we wanted to work with a linear recovery polynomial, we could stop here and set $p_i = \pi_{T_i^N}$.

Otherwise, if $T_i^{\min}$ must be found, we consider three sets of three triangles in the vicinity of $T_i^{\min}$, as defined in Fig. 10.5. For each of the three sets, we compute the quadratic polynomial

$$\sum_{|\alpha|\le 2} (x - c_i)^{\alpha} a_{\alpha}$$

from the condition $\frac{1}{\sigma_l} \int_{\sigma_l} \sum_{|\alpha|\le 2} (x - c_i)^{\alpha} a_{\alpha} = v_l$, where l runs over the indices belonging to the nodes of the three triangles in the set under consideration. In each case, this gives five equations for the five unknown coefficients. The sixth coefficient, $a_{(0,0)}$, is obviously the cell average v_i. From the three quadratic polynomials, we choose the one for which $\sum_{|\alpha|=2} |a_{\alpha}|$ is the smallest as the recovery polynomial.

We could now go on to enlarge the set of triangles, compute third-order polynomials, and so forth. However, the condition number for the linear systems gets worse very quickly. Finite-volume schemes with orders that are higher than three are therefore not found in the literature, except in the spatially one-dimensional case.

10.5
Remarks Concerning Non-polynomial Recovery

The only reason for using polynomials in the recovery process is their simplicity. Therefore, other choices for recovery functions can be found in the literature. One starting point when choosing recovery functions is the theory of optimal recovery, as founded by Golomb and Weinberger[161] and exploited by Micchelli and Rivlin.[162] The theory applies in particular to reproducing kernel Hilbert spaces in which the point evaluation functional δ_x, defined by means of

$$\delta_x u = u(\boldsymbol{x}) ,$$

is continuous. It can, however, be applied to arbitrary Banach spaces in which the point evaluation functional is continuous. In finite-volume methods, we actually need to recover point values at the Gaussian points of the edges between neighboring boxes. It turns out that we can find optimal functions in such spaces in the following sense.

Consider the recovery problem on box σ_i, and assume that we use $M-1$ additional boxes $\sigma_{i_1}, \sigma_{i_2}, \dots, \sigma_{i_{M-1}}$ in the vicinity of σ_i, so that the average data from M cells can be used. The given data consist of the cell averages for the M boxes. Let the space we want to explore be a Banach space V with a continuous point evaluation functional. We further assume the existence of a convex neighborhood $U \subset V$ of 0 for which

$$\forall u \in V \exists c > 0 : \quad cu \in U$$

and

$$u \in U \Leftrightarrow -u \in U$$

holds. We shall come back to the meaning of this convex neighborhood later. The mapping

$$V \ni u \mapsto \mathcal{I}_i u = \left(\frac{1}{\sigma_i} \int_{\sigma_i} u \, \mathrm{d}\sigma_i, \frac{1}{\sigma_{i_1}} \int_{\sigma_{i_2}} u \, \mathrm{d}\sigma_{i_2}, \dots, \frac{1}{\sigma_{i_{M-1}}} \int_{\sigma_{i_{M-1}}} u \, \mathrm{d}\sigma_{i_{M-1}} \right) \in \mathbb{R}^M$$

is called an *information operator*. The linear functional

$$V \ni u \mapsto \delta_x u = u(\boldsymbol{x})$$

is called a *property operator*. Every (not necessarily linear) operator

$$\mathbb{R}^M \ni \mathcal{I}_i u \mapsto \mathcal{R}_i(\boldsymbol{x}) \mathcal{I}_i u \in \mathbb{R}$$

161) M. Golomb, H.F. Weinberger: Optimal approximation and error bounds. In: R.E. Langer (ed.): *On numerical approximation*. Madison: Univ. Wisconsin Press 1959

162) C.A. Micchelli, T.J. Rivlin: Lectures on optimal recovery. In: P.R. Turner (ed.): *Numerical analysis, Lancaster 1984 (Lecture Notes in Mathematics 1129:12–93)*. Berlin: Springer 1984

is a *recovery operator*. Since we want to recover the point values of a function u, we define the *recovery error* by

$$E_{\mathcal{R}_i(x)}(\delta_x, \mathcal{I}_i, U) = \sup_{u \in U} |\delta_x u - \mathcal{R}_i(\boldsymbol{x})\mathcal{I}_i u| \,.$$

The lower bound

$$E(\delta_x, \mathcal{I}_i, U) = \inf_{\mathcal{R}_i(\boldsymbol{x})} E_{\mathcal{R}_i(x)}(\delta_x, \mathcal{I}_i, U) = \inf_{\mathcal{R}_i(\boldsymbol{x})} \sup_{u \in U} |\delta_x u - \mathcal{R}_i(\boldsymbol{x})\mathcal{I}_i u|$$

is called the *proper recovery error*. A recovery operator $\mathcal{R}_i^*(\boldsymbol{x})$ with the property

$$E_{\mathcal{R}_i^*(x)}(\delta_x, \mathcal{I}_i, U) = E(\delta_x, \mathcal{I}_i, U)$$

is an *optimal recovery operator*.

We have not used the convex neighborhood U around $0 \in V$, but it does have an important meaning. The problem of recovery from cell averages is not solvable in the form in which we stated it. This can be seen as follows. Consider all of the linear functionals contained in $\mathcal{I}_i u$ to be linearly independent and linearly independent from the property operator. Under these assumptions, there is a function $w \in V$ with

$$\delta_x w = 1$$

and

$$\forall k \in \{i, i_1, \ldots, i_{M-1}\} : \quad \frac{1}{\sigma_k} \int_{\sigma_k} u \, \mathrm{d}\sigma_k = 0 \,,$$

otherwise the hyperplane defined by $\delta_x w = 1$ would be parallel to one of the hyperplanes defined by the linear functionals in the information. The function $u + \xi w$ is an element of V for every u and $\xi \in \mathbb{R}$, and it satisfies the equation

$$\mathcal{I}_i(u + \xi w) = \mathcal{I}_i u + \xi \mathcal{I}_i w = \mathcal{I}_i u \,.$$

On the other hand, we have

$$\delta_x(u + \xi w) = \delta_x U + \xi \delta_x w = u(\boldsymbol{x}) + \xi \,.$$

Thus, arbitrary values of ξ result in arbitrary values for δ_x.

In order to remedy this situation, the convex neighborhood U of $0 \in V$ comes into play. We need a linear *restriction operator* $\mathcal{T}$, defined by

$$V \ni u \mapsto \mathcal{T} u \in V \,.$$

Aided by this operator, we can define

$$U = \{u \in V \mid \|\mathcal{T} u\|_V \leq 1\} \,,$$

and it is then easy to show that U is the convex neighborhood sought. If we now ask for recovery in U instead of V, we do not encounter the problems described above any more, and we may be able to find an optimal recovery function. We refer the reader to monograph [20], in which the theory of optimal recovery is fully developed for finite volume methods.

The interesting case of reproducing kernel Hilbert spaces reveals that splines are the optimal recovery functions. A *spline* is defined as a function that minimizes the norm (or semi-norm) in a Hilbert space. It is in this context that *radial basis functions* play an important role, like the *thin-plate spline* in the Beppo–Levi space

$$\Phi(x) = \sum_{j=0}^{M-1} \lambda_j \frac{1}{\sigma_{i_j}} \int_{\sigma_{i_j}} |x - y|^{2(m-1)} \log(|x - y|) \, d\sigma_{i_j}^y + v(x) ,$$

where $i_0 = i$ and v is any function from the kernel of the semi-norm in the Beppo–Levi space. For details on the theory and implementation of radial basis functions, see [20].

10.6 Remarks Concerning Grid Generation

We cannot provide an overview of grid generation techniques for unstructured grids here; the reader should instead refer to the literature available on the subject.[163] In the case of triangulations in 2D, there are very useful *adaptation algorithms* for refining or coarsening a given grid. We employ two different strategies to refine a triangle, namely *red refinement* and *green refinement*, as shown in Fig. 10.6.

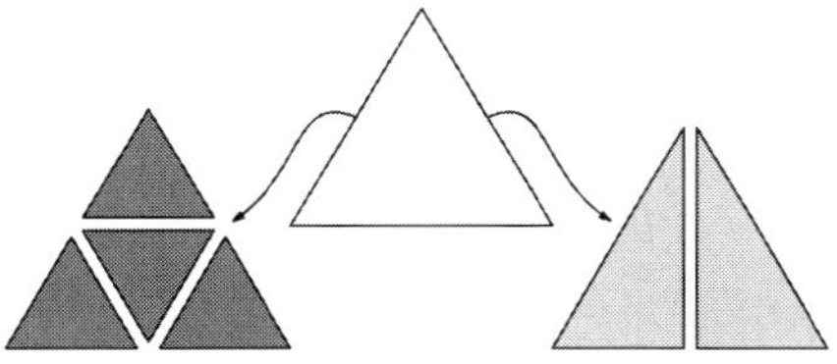

Fig. 10.6 Red (*left*) and green (*right*) refinement. (Courtesy of Cambridge University Press)[164]

Red refinements are preferable because they result in geometrically similar, smaller triangles, while green refinement leads to small angles. Hence, green refinement must only be used to fit a region that was red-refined to coarser triangulation in order to avoid hanging nodes.

If certain patches of triangles around a node have been refined they can also be recoarsened. Figure 10.7 shows a so-called resolvable patch which can be coarsened.

Another resolvable patch is shown in Fig. 10.8. The complete algorithms, along with many examples of their use, are given in the literature.[165]

163) See, for example, J.F. Thompson, B.K. Soni, N.P. Weatherill (eds.): *Handbook of grid generation*. Boca Raton: CRC 1999

164) Figures 10.6–10.12 are taken from K.W. Morton, T. Sonar: Acta Num. 16 (2007) 155–238

165) See, for example, Chapter 3 in [15]

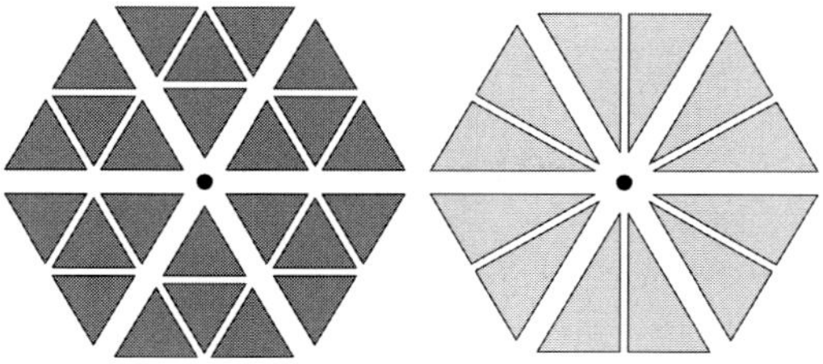

Fig. 10.7 Resolvable patch (*left*) and recoarsened patch (*right*). (Courtesy of Cambridge University Press)

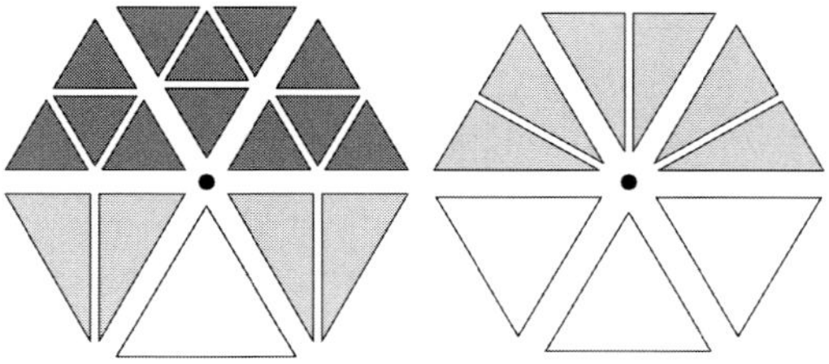

Fig. 10.8 Resolvable patch (*left*) and recoarsened patch (*right*). (Courtesy of Cambridge University Press)

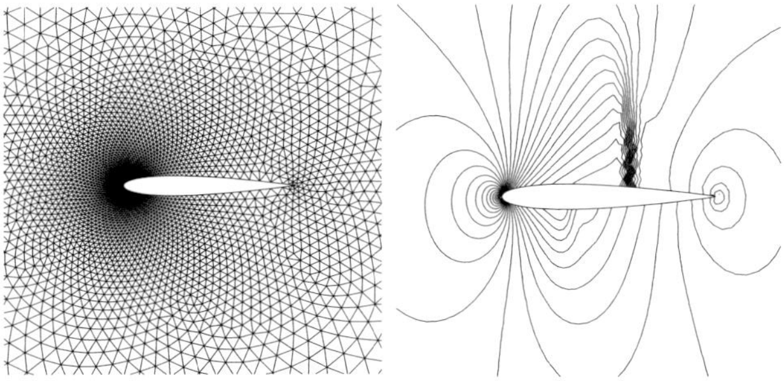

Fig. 10.9 Grid and pressure distribution. (Courtesy of Cambridge University Press)

In Fig. 10.9, we show a grid in the vicinity of a NACA 0012 airfoil and the pressure distribution computed using a second-order finite-volume method. The angle of attack is $\alpha = 1.25^\circ$ and the onflow Mach number $\mathrm{Ma}_\infty = 0.8$. Under these conditions, we expect a strong shock on the upper surface of the airfoil and a weak one on the lower side.

Both shocks can be seen in the pressure distribution but they are smeared out. If an *error indicator* is used to mark the triangles to be refined, we arrive at a refined grid such as that shown in Fig. 10.10. The corresponding pressure distribution now shows nicely captured shocks.

A very impressive case is shown in Fig. 10.11. Here, an inviscid flow with $\mathrm{Ma}_\infty = 3$ arrives from the left and hits a forward-pointing step in a channel. The resulting flow in Fig. 10.12 shows a nice pattern of shocks, a contact discontinuity and a rarefaction wave.

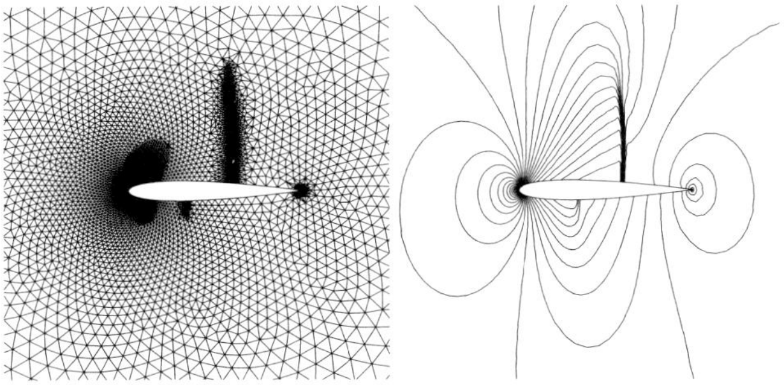

Fig. 10.10 Grid and pressure distribution. (Courtesy of Cambridge University Press)

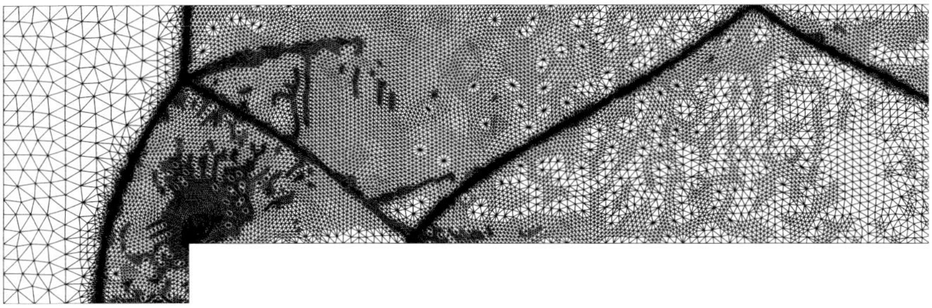

Fig. 10.11 Adapted grid for channel flow. (Courtesy of Cambridge University Press)

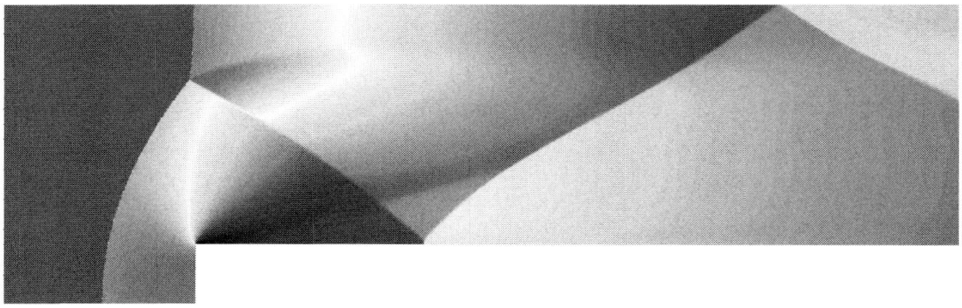

Fig. 10.12 Density distribution of channel flow. (Courtesy of Cambridge University Press)

One question that we are yet to address is the selection of an error indicator. In order to get true control over the error, we need reliable information about the function space of solutions of the Euler or Navier–Stokes equations. This is presently not the case. However, a lot of work has recently been done on hyperbolic systems, and this can be found in the literature.[166)]

166) See, e.g., P. Houston, J. Mackenzie, E. Süli, G. Warnecke: Num. Math. **82** (1999) 433–470; T. Sonar, E. Süli: Num. Math. **78** (1998) 619–658

Index

a

adiabatic exponent 6
advection equation 13
amplification factor 142
amplification matrix 143
angle of attack 18
angular velocity vector 15
antidifference operator 134
approximate solutions 116
Archimedes 30
Archimedes' principle 30
artificial time 90
asymptotically closed 117
asymtotically regular 117
Aw, A. 77
Aw–Rascle model 77

b

back side of a shock 45
basic discretization 208
Bernoulli equation 16
– generalized 17
Bernoulli, Daniel 16
Bois-Reymond, P. du 119
Boltzmann constant 106
Boltzmann, L. 49
boundary layer 95
boundary layer equations 99
boundary layer theory 89
box 206
Brandt, G.B. 106
Breuss, M. 71
Briggs, H. 131
Buchanan, J.E. 144
Buckley–Leverett equation 71
buoyancy
– dynamic 14
– static 26, 30
Burgers' equation 10
Burgers, J. 10

c

carburetor 16
Cauchy–Riemann equations 20
cell average 130
centered waves 63
CFL condition 114, 122, 136, 140, 159
– stronger 161
Chapligin solutions 28
Chapligin, C.A. 28
characteristic cone 76
characteristic field 170
– genuinely nonlinear 170
– linearly degenerated 170
characteristics 10–12, 41
– numerical 135
circulation 15
Clausius, R.J.E. 49
closing of disturbed laminar flow 102
congestion 38
congestion velocity 47
conservation laws 7
conservation of
– energy 1, 5
– mass 1, 3
– momentum 1, 3
conservation principles xv, 1
conservative variables 5
consistency condition 138
consistency of a numerical method 122
consistency of numerical methods 133
contact discontinuity 45, 171
continuity equation 3
continuous convergence 119
– of relations 151
contravariant components of a vector 190
contravariant metric tensor 192

Mathematical Models of Fluid Dynamics. R. Ansorge and T. Sonar

ISBN: 978-3-527-40774-3

control volume 206
convection term 5
convergence of numerical methods 74, 117
Courant, R. 114, 119, 135
covariant basis vectors 190
covariant components of a vector 191
covariant metric tensor 192
critical Reynolds number 89, 104
curve of neutral stability 104

d
density 1, 7
density of momentum 3
difference equation 128
difference operator 127, 138
– central 132
– forward 131
difference stencil 135
diffusion term 14
discontinuity
– admissible 170
discontinuous Galerkin method 129
discrete entropy conditions 119
discrete Fourier transform 141
discretely compact 117
divergence theorem 4
drag force 85

e
e^9-rule 104
elements of FEM 128
elliptic grid generation 202
energy
– interior 5
– kinetic 5
– total 5
Engquist, B. 124
Engquist–Osher scheme 124
ENO schemes 148
entropy 7, 49
entropy condition 50, 52
entropy flux 53, 121
– numerical 123
entropy functional 53
entropy functions 121
entropy inequalities 68
entropy inequality 163
entropy solution 50, 53, 57, 62, 67, 68, 70, 118, 121, 163
– discrete 119, 123
equation of state 2, 7
equivalence theorem 139
Euler equations 1, 5
– entropy formulation 169
Euler's law of vanishing momentum 80
Euler, L. 5, 80, 131
evolution equation 137
evolution equation for cell averages 130
evolution operator 137
– discrete 138
exact difference formula 132
existence proof for entropy solutions 113
explicit difference equation 128

f
Fellehner, St. 91
finite difference grid 128
finite difference method (FDM) 73, 127
finite element method (FEM) 127
finite volume method (FVM) 127
First law of thermodynamics 5
flow
– barotropic 16
– incompressible 7
– irrotational 16
– isentropic 52
– laminar 85
– nonstationary 8
– plane 18
– potential 14, 15
– similar 85
– stationary 8
– steady-state 8
– supersonic 18
– through porous media 71
– turbulent 85
– subsonic 18
flow separation 95
fluid
– Newtonian 81
– ideal 4
– inviscid 4
– real 4
– viscous 4
flux 7
– numerical 122
flux function
– numerical 130
flux limiter method 147
flux splitting 123

flux splitting scheme 124
force
  – conservative 16
Friedrichs, K.-O. 114, 135
front side of a shock 45
functional analysis of discretization algorithms 115
fundamental diagram 38
FVM on unstructured grids 148

g
gas
  – γ- 7
  – ideal 4
Gasser, I. 106
genuine solution 137
genuinely nonlinear 63
Glimm, J. 114
global discretization error 140
Godunov method
  – for Euler equation 185
Godunov scheme 160
Godunov, S.K. 74
Greenshields' model 39
grid 73
grid generation 149, 196
grid generation algebraic 198
grid generation analytic 198
grid generation cycle 202
grid of C-type 196
grid of H-type 197
grid of O-type 197
grid quality 197
grid topology 196
grid transformation 188

h
Hagen, G.H.L. 89
Hagen-Poiseuille law 88
Harriot, T. 131
Harten, A. 121, 148
heated gas flow 104
Hilgenstock 201
hodograph equation 28
hodograph plane 27
hydrodynamic paradoxon 16
hydrostatic equation 29
hydrostatic paradox 29
Hyman, J.M. 121

i
implicit difference equation 128
incremental form 157
infinitesimal generator 137
iterative procedure 91

j
Jacobian matrix 7
Jagger, S.F. 106
jam concentration 38
jet stream 16
jump condition 44

k
Kantorowitch, L.W. 114
Kelvin's theorem 17
Kelvin, Lord of Largs 17
Keyfitz, B. 125
Kreiss matrix theorem 144
Kreiss, H.-O. 144
Kruzkov solution 70
Kruzkov, S.N. 69, 121
Kutta, M.W. 26
Kutta–Zhukovsky buoyancy formulas 26
Kuznetsov, N.N. 59

l
Lahmann, J.-R. 104
Laurent, P.A. 23
Lax entropy solution 55
Lax shock conditions 68
Lax, P. 53, 121
Lax–Friedrichs scheme 155
Lax–Richtmyer theory 136
Lax–Wendroff scheme 147
Lax-Friedrichs method 164
Lax-Friedrichs scheme 146
Lax-Richtmyer theory 139
Lax-Wendroff theorem 125
Lea, C.J. 106
Legendre transformation 27
Lei, J. 126
Leibniz, G.W. 131
LeVeque, R.J. 125
Lewy, H. 114, 135
lift 26
Lighthill, M.J. 39
Lighthill–Whitham model 39, 77
limiter 148
local speed of sound 32

m
Mach number 18, 176
Mach's angle 19
Mach's cone 19

Mach's net 19
Mach, E. 18
main theorem of thermodynamics
– first 5, 49
– second 49
material time derivative 5
mathematical model xiii
maximum principle 158
method of
– characteristics 19
– hodographs 27
method of artificial compressibility 90
method of conservation form 122, 146
modified equation 144
monotone schemes 124
monotonicity preserving 156
Morton, K.W. 144
multistep method 74
Murman–Courant–Isaacson–Rees method 160

n
Napier, J. 131
Navier, C.L.M.H. 83
Navier–Stokes equations 83
Neumann, J. von 141
Newton's second law 3
Nickel, K. 95
no-slip condition 26, 106
non-standard conservation law 71
numerical domain of dependence 135
numerical flux function 155
numerical viscosity coefficient 156

o
Oleinik solution 58
Oleinik, O. 58
one-step method 74
optimal recovery 216
order of a method 138
order of a numerical procedure 133
Orr–Sommerfeld equation 103
Osher, St. 124

p
Panov, E.Y. 70
parameter space 187
partial slip 26
Petrov–Galerkin schemes 129
phase speed of perturbations 104
physical space 187
Plum, M. 104
Poise 83
Poiseuille, J.L.M. 89
polar coordinates 187
polymer 72
Prandtl number 85
Prandtl, L. 89, 95
pressure 6
– dynamic 16
– static 16
– total 24
primary grid 206
primitive variables 5
pseudospectral method 131

q
quasilinear systems of partial differential equations 7

r
Rankine–Hugoniot condition 42, 44, 170
Rankine-Hugoniot condition 76
rarefaction wave 41, 68, 172
Rascle, M. 77
reconstruction of functions 74
recovery polynomial 211
Reynolds averaging 102
Reynolds number 84
– critical 89, 104
Reynolds' transport theorem 2, 104
Reynolds, O. 2
Richtmyer, R.D. 114, 144
ride impulse 41
Riemann invariant 68
Riemann invariants 170
Riemann problem 39, 73
– complete solution 178
– local 160
Riemann solvers 74
Riemann, B. 39
Rinow, W. 120
round-off errors 140

s
scaled quantities 107
scheme
– (2k+1)-point 122, 146
– 3-point 135, 146
– upwind 136
Schlichting, H. 97
second law of thermodynamics 49

secondary grid 206
semigroup property 137
shallow water equations 35
shift operator 131
shift operator
– backward 141
– forward 142
shock 42, 45, 68, 171
shock conditions 68
shock curve 45
shock Mach number 176
similarity solution 171
slope limiting 214
smooth connection of states 66
solution operator 137
– generalized 137
Sommerfeld, A. 103
sound propagation 29
sound velocity, local 9, 32
source terms 200
spatial order of a FVM 210
specific heats 7
specific thermal conductivity 85
spectral collocation method 131
spectral discretizations 131
spectral element method 131
spectral Galerkin method 131
stability
– numerical 114, 139
– of laminar flows 102
– spatial 103
– temporal 103
stability criterions 143
step size 74
Stokes approximation 89
Stokes, G.G. 83
stream function 20
streamline 9
streamline diffusion methods 129
stress tensor 80
strictly hyperbolic 9
Struckmeier, J. 106
Stummel, F. 114
sucking off of the boundary layer 99
summation by parts 134
symbol of a shift operator 141

t
Tadmor's lemma 165
temperature 7
test elements 70, 116
test function 35
test functions 128
thermal conductivity 105
time step size 73, 122
Tollmien, W. 95
Tollmien-Schlichting waves 103
traffic flow 37
traffic jam 46
transformed divergence term 195
transformed Laplace equation 195
transport theorem 2
tribology 89
truncation error 74, 133, 138
tunnel fires 106
TVB schemes 124
TVD property 156
TVD schemes 124
Tveito, A. 71

u
unscaled quantities 107
unstructured grids 130, 148, 218

v
Vainikko, G. 115
variables
– conservative 5
– primitive 5
velocity potential 15
– complex 20
viscosity 4, 72
– kinematic 83
viscosity coefficient 155
– first 83
– second 83
viscosity form 156
von Neumann stability criterion 144
vortex line 15
vorticity vector 15

w
Wang, C.Y. 89
weak formulation of a numerical method 116
weak solution 13, 35, 67, 116
Wendroff, B. 125, 147
Whitham, C.B. 39
wing theory 89
Winther, R. 71

z
Zhukovsky, N.J. 26

Suggested Reading

The aim of this book is to provide an introduction to all of the aspects of mathematical fluid dynamics (MFD); it would be impossible to cover all of these aspects in depth in this one book. However, many excellent, modern and extensive monographs on the subject are available. Most of them describe particular areas of MFD (like numerical procedures), they treat the basic equations theoretically, or they justify the mathematical treatment from a physical or engineering point of view. Some modern books that are useful for more detailed studies are listed here:

1 Brower W.B., *Dynamics of flows in one space dimension*, Boca Raton: CRC, **1999**

2 Chorin A.J., Marsden J.E., *A mathematical introduction to fluid mechanics, 2nd ed*, New York: Springer, **1984**

3 Dubois T., Jauberteau F., Temam R., *Dynamic multilevel methods and the numerical simulation of turbulence*, Cambridge: Cambridge Univ., Press, **1999**

4 Emanuel G., *Analytic fluid dynamics*, Boca Raton: CRC, **2001**

5 Feistauer M., *Mathematical methods in fluid dynamics*, Boca Raton: CRC, **1993**

6 Ferziger J.H., Peric M., *Computational methods for fluid dynamics*, Berlin: Springer, **1999**

7 Freistühler H. (ed), *Analysis of systems of conservation laws*, Boca Raton: Chapman&Hall/CRC, **1998**

8 Frisch U., *Turbulence: the legacy of A.N. Kolmogoroff*, Cambridge: Cambridge Univ. Press, **1995**

9 Godlewski E., Raviart P.-A., *Numerical approximation of hyperbolic systems of conservation laws*, Berlin: Springer, **1996**

10 Holmes P., Lumley J.L., Berkooz G., *Turbulence, coherent structures, dynamical systems and symmetry*, Cambridge: Cambridge Univ. Press, **1996**

11 Kröner D., *Numerical schemes for conservation laws*, Chichester: Wiley Teubner, **1997**

12 LeVeque R.J., *Numerical methods for conservation laws*, Basel: Birkhäuser, **1990**

13 Majda A., *Compressible fluid flow and systems of conservation laws in several space variables*, Berlin: Springer, **1984**

14 Malek J., Necas J., Rokyta M., *Advanced topics in theoretical fluid mechanics*, Boca Raton: CRC, **1999**

15 Meister A., Struckmeier J., *Hyperbolic partial differential equations*, Braunschweig: Friedr. Vieweg, **2002**

16 Richtmyer R.D., Morton K.W., *Difference methods for initial value problems, 2nd ed*, New York: Wiley, **1967**

17 Schlichting H., Gersten K., *Boundary-layer theory, 8th ed*, Berlin: Springer, **2000**

18 Serre D., *Systems of conservation laws 1*, Cambridge: Cambridge Univ. Press, **1999**

19 Smoller J., *Shock waves and reaction-diffusion equations*, Berlin: Springer, **1983**

20 Sonar T., *Mehrdimensionale ENO-Verfahren*, Stuttgart: B.G. Teubner, **1997**

21 Spurk, J.H., *Fluid mechanics*, Berlin: Springer, **1997**

22 Warsi, Z.U.A., *Fluid dynamics, theoretical and computational approaches*, Boca Raton: CRC, **1998**

Mathematical Models of Fluid Dynamics. R. Ansorge and T. Sonar

ISBN: 978-3-527-40774-3